Large-Eddy Simulation in Hydraulics

T0172777

IAHR Monograph

Series editor

Peter A. Davies
Department of Civil Engineering,
The University of Dundee,
Dundee,
United Kingdom

The International Association for Hydro-Environment Engineering and Research (IAHR), founded in 1935, is a worldwide independent organisation of engineers and water specialists working in fields related to hydraulics and its practical application. Activities range from river and maritime hydraulics to water resources development and eco-hydraulics, through to ice engineering, hydroinformatics and continuing education and training. IAHR stimulates and promotes both research and its application, and, by doing so, strives to contribute to sustainable development, the optimisation of world water resources management and industrial flow processes. IAHR accomplishes its goals by a wide variety of member activities including: the establishment of working groups, congresses, specialty conferences, workshops, short courses; the commissioning and publication of journals, monographs and edited conference proceedings; involvement in international programmes such as UNESCO, WMO, IDNDR, GWP, ICSU, The World Water Forum; and by co-operation with other water-related (inter)national organisations. www.iahr.org

Supported by
CEDEX

Large-Eddy Simulation in Hydraulics

Wolfgang Rodi
Karlsruhe Institute of Technology (KIT), Germany/
King Abdulaziz University, Jeddah, Saudi Arabia

George Constantinescu
The University of Iowa, USA

Thorsten Stoesser
Cardiff University, UK

CRC Press
Taylor & Francis Group
Boca Raton London New York Leiden

CRC Press is an imprint of the
Taylor & Francis Group, an **informa** business

A BALKEMA BOOK

CRC Press/Balkema is an imprint of the Taylor & Francis Group, an informa business

© 2013 Taylor & Francis Group, London, UK

Typeset by V Publishing Solutions Pvt Ltd., Chennai, India
Printed and Bound by CPI Group (UK) Ltd, Croydon, CR0 4YY

Published by: CRC Press/Balkema
P.O. Box 11320, 2301 EH Leiden, The Netherlands
e-mail: Pub.NL@taylorandfrancis.com
www.crcpress.com – www.taylorandfrancis.com

Library of Congress Cataloging-in-Publication Data

Applied for

ISBN: 978-1-138-00024-7 (Hbk)
ISBN: 978-0-367-57638-7 (Pbk)
ISBN: 978-0-203-79757-0 (eBook)

About the IAHR Book Series

An important function of any large international organisation representing the research, educational and practical components of its wide and varied membership is to disseminate the best elements of its discipline through learned works, specialised research publications and timely reviews. IAHR is particularly well-served in this regard by its flagship journals and by the extensive and wide body of substantive historical and reflective books that have been published through its auspices over the years. The IAHR Book Series is an initiative of IAHR, in partnership with CRC Press/ Balkema – Taylor & Francis Group, aimed at presenting the state-of-the-art in themes relating to all areas of hydro-environment engineering and research.

The Book Series will assist researchers and professionals working in research and practice by bridging the knowledge gap and by improving knowledge transfer among groups involved in research, education and development. This Book Series includes Design Manuals and Monographs. The Design Manuals contain practical works, theory applied to practice based on multi-authors' work; the Monographs cover reference works, theoretical and state of the art works.

The first and one of the most successful IAHR publications was the influential book *"Turbulence Models and their Application in Hydraulics"* by W. Rodi, first published in 1984 by Balkema. I. Nezu's book *"Turbulence in Open Channel Flows"*, also published by Balkema (in 1993), had an important impact on the field and, during the period 2000–2010, further authoritative texts (published directly by IAHR) included *Fluvial Hydraulics* by S. Yalin and A. Da Silva and *Hydraulicians in Europe* by W. Hager. All of these publications continue to strengthen the reach of IAHR and to serve as important intellectual reference points for the Association.

Since 2011, the Book Series is once again a partnership between CRC Press/ Balkema – Taylor & Francis Group and the Technical Committees of IAHR and I look forward to helping bring to the global hydro-environment engineering and research community an exciting set of reference books that showcase the expertise within IAHR.

Peter A. Davies
University of Dundee, UK
(Series Editor)

Table of contents

Preface

Numerical computation methods are used more and more as tools for solving hydraulic and environmental engineering problems, and as turbulence mostly plays an important role in these problems, the realistic simulation of the effect of turbulence is essential. 34 years ago the first author published a book on Turbulence Models and their Application in Hydraulics in the IAHR Monograph Series. The statistical models described there, now known as RANS models, have to account for the entire effect of the complex turbulent motions since methods relying on them do not resolve the turbulent fluctuations. Such models became the workhorse for calculating turbulent flows in hydraulics. In the meantime, the computer power has increased tremendously, allowing to calculate more and more complex problems, which approach real-life situations. This has revealed the limitations of the statistical RANS models, especially in situations when large-scale turbulent structures dominate the flow and mixing behaviour. The increase in computer power has also led to the development of more powerful but computationally more demanding simulation methods for turbulent flows, moving away from purely statistical treatment to eddy-resolving techniques. These can take much better account of the physics of the complex turbulent motion and allow also to study and help understand the mechanisms involved. As Direct Numerical Simulation (DNS) resolving eddies of all sizes is out of reach for practical applications, the development concentrated mainly on Large-Eddy Simulations (LES) in which, as the name implies, only the large scales are resolved while the effect of the small scales is modelled. In the last 2 decades the research has shifted clearly from RANS to LES methods, and as the latter became mature and the computer power kept increasing, they also became suitable tools for practical applications and hence are now included as modules in most commercial and open-source CFD software. As a consequence, LES is used increasingly not only in research but also for practical calculations in all fields of engineering, including hydraulics, and in geophysical applications.

Because of this shift in available and now often preferred methods and because of the great potential and future of LES, IAHR – through its then Vice President Prof. J.H.W. Lee – initiated the writing of a monograph on this method for the IAHR series as a follow-up to the now dated book in the series on RANS models. The new book was to be geared for hydraulic and environmental engineers, and hence the purpose of the book is to provide an easily understandable introduction to the

LES method for these engineers and to demonstrate to them the great potential of the method in their field of work. Note that for brevity only the word "Hydraulics" was retained in the title of the book and also in the main body it always includes environmental fluid mechanics, albeit restricted to the wet environment. A number of books on and reviews of LES are available already and are listed in the Introduction (Section 1.4), but these are not geared on hydraulics and are mostly too theoretical for the practicing engineer. The present book is intentionally less theoretical and mathematically demanding and hence, hopefully, easy to follow. Also, it covers special features of flows in water bodies and summarizes the experience gained with LES for calculating such flows. However, much of what is covered in the book is also useful to readers working in other areas of fluid mechanics where turbulence plays an important role.

The book introduces first the basic methodology of LES and then the various sub-components of the method such as subgrid-scale models, near-wall models, boundary condition treatment, numerical methods and structure-education methods. The large variety of approaches and models for these proposed in the literature cannot all be included and described in detail in the book. Only the most commonly used and sustained ones are covered, in particular those successfully applied in hydraulics. For some methods/models only a brief summary is given and the reader is referred to the literature for a detailed description. This applies for example to the Implicit LES method (ILES) not employing an explicit subgrid-scale model, which has recently gained popularity, albeit not so much in hydraulics applications. Substantial coverage is given to Hybrid RANS-LES methods, in which only part of the flow domain is treated by LES but another part by a much less expensive RANS method. This approach is particularly important for practical problems where the Reynolds number is high and/or the flow domain is large.

The power and potential of the LES method is demonstrated in an extensive chapter presenting application examples of hydraulic flows, ranging from simple straight open channel flow to complex situations of various kinds. Most examples stem from the research groups of the authors because these were best known and available to them. Hence, there is a certain bias towards the authors' calculations, but results of others are also presented. The majority of examples are pure flow calculations, but some simulations with mass transfer are also included. However, the examples are restricted to single-phase flow situations. LES is also used increasingly for sediment-transport calculations, but such applications are not presented as they involve the use of additional models for this transport which are not covered in this book.

The suggestion to write this monograph for the IAHR Series came from Prof. J.H.W. Lee in his capacity as Vice President of the IAHR. We should like to thank him for this stimulus and for his constant support. We are also grateful to the editor of the IAHR Monograph Series, Prof. P.A. Davies, for his valuable advice and for facilitating the publication of this book. We have benefited over the years from fruitful discussions and information exchange with Profs. M. Breuer and J. Fröhlich as well as Dr. D. von Terzi, which we should like to acknowledge. We thank Dr. S. Hickel for his comments on and input to a draft of Chapter 5 on ILES and Prof. U. Piomelli for providing access to his dune-flow LES results. We also extend special thanks to our numerous Ph.D. students and post-doctoral workers for their research contributions which form the basis of many of the examples in the applications section. We are grateful to our home institutions for the infrastructure provided to us and to the various funding agencies that have supported our research in LES over the years. G. Constantinescu completed part of his contribution to this book while on sabbatical leave at the laboratory of Hydraulics, Hydrology and Glaciology (VAW) at ETH Zurich, Switzerland. He would like to thank Prof. W. Hager and the other researchers at VAW for their support. Finally, we thank Janjaap Blom and Lukas Goosen of CRC Press/Balkema for the good cooperation in the preparation of this book.

Wolfgang Rodi
Karlsruhe
George Constantinescu
Iowa City
Thorsten Stoesser
Cardiff
March 2013

Chapter 1

Introduction

As in other fields of engineering, in hydraulic engineering numerical computation methods are used more and more for studying the physical phenomena involved and for solving practical problems. Here, and in fact in the entire book, hydraulic engineering always includes environmental engineering and hydraulics includes environmental fluid mechanics, albeit restricted to the wet environment. With the increasing computer power and the availability of advanced numerical methods, increasingly more complex and practically realistic problems can be solved by numerical methods. A difficulty in obtaining reliable predictions is the fact that almost all flows in hydraulics, whether geophysical or man-made flows, are turbulent, except for groundwater flows. The unsteady irregular eddying motion associated with turbulence increases greatly the momentum, heat and mass transfer and hence has significant influence on all aspects of the flow and associated phenomena such as temperature, concentration distributions and sediment transport etc., as will shortly be outlined in some detail. Hence, in any successful computation of flow and associated phenomena, a realistic simulation of the effect of turbulence is important. This book deals with an advanced, powerful method for simulating turbulence, namely the Large Eddy Simulation (LES), which, also for practical problems in hydraulics, can be seen as the method of the future.

1.1 THE ROLE AND IMPORTANCE OF TURBULENCE IN HYDRAULICS

In order to emphasize the need for simulating realistically turbulence and its effect, the importance and role that turbulence plays in hydraulic engineering problems is briefly outlined first. The increase in momentum transfer caused by the fluctuating turbulent motion increases the friction on solid boundaries of the flow and as a consequence causes losses in flows through conduits and around structures. It thereby determines the flow rate and pressure drop in conduits and the water level in open channels, and it also determines the rate of energy dissipation.

Turbulence has also a governing influence on the details of the flow development, such as the velocity distribution, the pressure distribution in the flow and along its boundaries and hence the forces on structures, including unsteady forces. For example, turbulence causes the velocity distribution in pipe flow to be much more uniform than in laminar pipe flow (parabolic distribution) and it is turbulence that causes secondary motions in non-circular conduits and open channels.

The mixing due to the fluctuating turbulent motion is responsible for the spreading of jets and the entrainment of ambient fluid and also the dilatation of discharged pollutants; and this mixing is also responsible for the washing out of substances such as pollutants from semi-enclosed regions like bays, harbours and groyne fields, and the turbulence also governs the conditions in tanks and basins of all kinds, whether there is flow through or stirring.

It is turbulence that keeps sediment particles in suspension, counteracting their settling by gravity, and it is also turbulence that erodes particles from a river bed. Hence, both bed-load and suspended-load sediment transport are governed by the turbulent motion.

Finally, as turbulence also causes and controls the gas exchange at the free surface between water bodies and the atmosphere, it controls aeration. It also controls flocculation and biological reactions within the water body and hence has a great influence on the water quality.

1.2 CHARACTERISTICS OF TURBULENCE

For an appreciation of the task and the difficulties of simulating turbulence and its effects in a calculation procedure, the characteristic features of turbulence are introduced and discussed first. An excellent introduction to the subject of turbulence and a visualization of its phenomena can be found in the old but by no means outdated film of Stewart (1969) which is available on the internet, and turbulence in open channel flow is treated comprehensively in the book of Nezu and Nakagawa (1993). Unfortunately, for the modeler, turbulent motions are very complex as they are fairly irregular, always unsteady and three-dimensional. This is in contrast to laminar flow which is regular and can be steady as well as two-dimensional or one-dimensional. In turbulence, flow quantities such as velocity, temperature or concentration undergo complex variations with space and time, manifested by fluctuations. These turbulent fluctuations cause strong momentum, heat and mass transfer and greatly increase the molecular transfer due to the Brownian motion in laminar flow. This increased transfer and the strong mixing caused by it is the practically most important feature of turbulence.

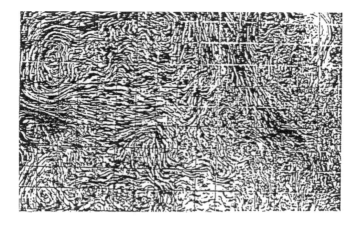

Figure 1.1 Turbulent eddies at the surface of a stirred water tank (courtesy of Reinhard Friedrich).

Turbulent motion carries vorticity, with rotation axes in all directions, and is composed of eddies interacting with each other, and an important characteristic of turbulence is that it consists of a spectrum of eddy sizes which is illustrated in Figure 1.1 showing a visualization of the eddies at the free surface of water in a stirred tank. The eddy sizes range from large eddies of the size of the flow domain, e.g. the pipe diameter or jet width, corresponding to low frequency fluctuations, to small eddies at which viscous forces act and dissipation takes place, corresponding to high frequency fluctuations. Figure 1.2 provides an example of a turbulent spectrum showing the distribution of the kinetic energy of the fluctuations with the wave number k of the turbulence which is proportional to the frequency but is the inverse of the eddy sizes. The large eddies are generally the most energetic ones and they extract energy from the mean motion; they break up into smaller eddies and transfer their fluctuating energy to the smaller eddies with higher frequency fluctuations, a process called energy cascade. This breaking-up and transfer to smaller and smaller eddies continues until viscous forces become active and the fluctuations are damped by these which is the process of dissipation of fluctuation energy occurring at the smallest eddies. In Figure 1.2, spectra are given for 3 Reynolds numbers and it can be seen that the width of the spectrum increases with the Reynolds number. This means that at high Reynolds numbers the dissipation takes place at smaller eddies and that the ratio of the sizes of smallest to large eddies increases with the Reynolds number. At sufficiently high Reynolds numbers there is a middle range in eddy sizes and wave numbers where energy is neither fed from the mean motion nor dissipated by viscous forces by the smallest eddies but only transferred from the larger to the smaller eddies. This is called the inertial sub-range indicated by T in the figure and here the spectrum behaves as $E \sim k^{-3/5}$. The width of this range increases also with the Reynolds number as can be seen from Figure 1.2.

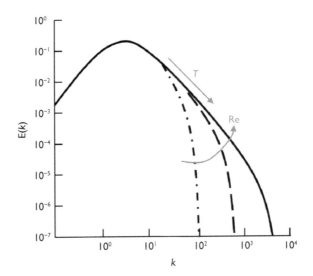

Figure 1.2 Spectra of isotropic turbulence with increasing Reynolds number $Re_t (= tke^{1/2} L/v, tke = $ kinetic energy and $L = $ macro length scale of turbulence), after Fröhlich (2006).

In shallow water flows, the situation can be different: Here the horizontal extent of the water body is much larger than the water depth and predominantly two-dimensional eddies with vertical axes of rotation can exist with sizes considerably larger than those of the bed-friction generated 3D turbulence which is restricted by the water depth. In this case, a two-range spectrum can exist as shown schematically in Figure 1.3. The low wave number range is the spectrum of the 2D eddies with horizontal motion and the higher wave number range represents the spectrum of the 3D eddies with size smaller than the water depth. In the range of 2D eddies, energy may be transferred from smaller to larger eddies so that in this part of the spectrum one can have an inverse cascade as indicated in the figure. This is however not genuine turbulence, yet the eddies are highly unsteady and contain considerable energy and hence have a strong influence on the transport of momentum and scalar quantities. For a successful calculation of shallow water flows, the effects of both ranges need to be simulated.

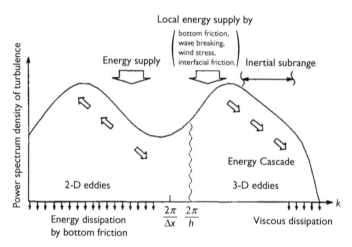

Figure 1.3 Two-range energy spectrum, schematic sketch from Nadaoka and Yagi (1998), reproduced with permission from ASCE.

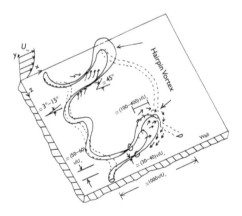

Figure 1.4 Schematic sketch of bursting phenomena, from Nezu and Nakagawa (1993).

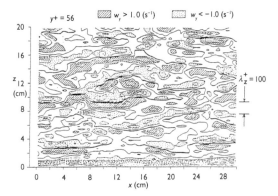

Figure 1.5 Streaky structures near bed of open-channel flow; measurements of vertical component of vorticity at y+ = 56 (from Nezu and Nakagawa, 1993).

Figure 1.6 Coherent structures in a shallow mixing layer, courtesy of W. Uijttewaal, see also Van Prooijen and Uijttewaal (2002).

The small-scale turbulent motions, where viscous forces act, and also those in the inertial sub-range behave fairly randomly, so that the fluctuations can be described by a Gaussian probability density function. The larger eddies interact with the mean flow and depend on the boundary conditions and hence the flow situation considered and are generally not independently random but often have some order and some correlated behaviour and are hence labeled as coherent. These coherent structures have a lifecycle, including birth, development and convection by the mean motion during which they retain their character, interaction between themselves and finally breakdown. A definition of what is a coherent structure is given in the book of Nezu and Nakagawa (1993) and there also the general behaviour of coherent structures is discussed at some length. Nezu and Nakagawa also give a classification of coherent structures and introduce two main groups, namely (i) Bursting phenomena near walls which involve injections and sweeps, hairpin vortices as well as low and high speed streaks; (ii) large-scale vortical motions away from walls which are induced by the mean flow and/or the geometry of the boundary. A schematic sketch of the bursting phenomena is given in Figure 1.4 and an example of the streaky structure of the turbulent fluctuations very near walls as extracted from experiments is shown

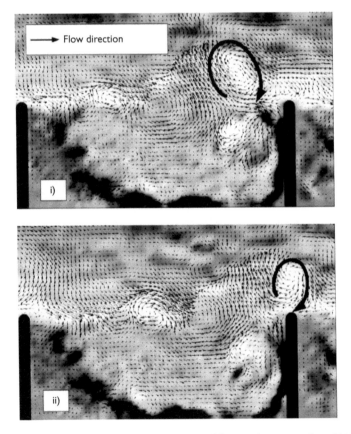

Figure 1.7 Measured coherent structures in open-channel flow with groynes, from Weitbrecht et al. (2008), reproduced with permission from ASCE.

in Figure 1.5. Examples of large-scale vortical structures away from walls are presented in Figures 1.6 and 1.7, Figure 1.6 showing structures in a mixing layer arising from the instability of the shear layer between streams with different velocity, and Figure 1.7 showing structures bordering the cavity of a groyne which in this case are geometry induced. Further examples of coherent structures are given in chapter 8 on Eduction Methods, where also a table summarizing various types of coherent structures is presented. In the eduction chapter various ways of detecting and visualizing the coherent structures are introduced. A special feature of the large-eddy simulation technique, in contrast to methods averaging out the turbulence, is that the method resolves the coherent structures and hence accounts directly for their often strong effect on the transfer of momentum, heat and mass.

1.3   CALCULATION APPROACHES FOR TURBULENT FLOWS

In this section, the three main approaches for simulating turbulent motions and their effects are briefly introduced. Turbulence is governed by the same basic equations as general laminar flows, namely the unsteady 3D Navier Stokes equations together with the continuity equation, and together these describe all the complex details of the turbulent fluctuating motion. In situations with heat or mass transfer, the corresponding scalar transport equation has to be added. Today, these equations can be solved numerically, and a method that does so without introducing any model is called Direct Numerical Simulation (DNS). In such an approach all scales of the turbulent motion from the large ones down to the smallest, dissipative scales must be resolved. Hence, the size of the numerical mesh must be smaller than the size of small-scale motions where dissipation takes place. In Figure 1.1 a mesh is indicated and it can be seen that eddies exist with a size smaller than the mesh size so that these could not be resolved on the given mesh. A much finer numerical grid would have to be used to obtain a resolution of all eddies in this turbulent flow. As the relation of the size of the smallest eddies to the size of the large-scale motion and hence the flow domain varies inversely with the Reynolds number (W.C. Reynolds 1990) and the calculations always have to be 3D, the number of grid points and the computing cost required increase roughly with Re^3. As a consequence, for Reynolds numbers of practical relevance the number of grid points required becomes so large that the computational effort required exceeds the capabilities of available computers. Even at medium Reynolds numbers and for simple flows, the computing effort is enormous. This is exemplified by the DNS of Hoyas and Jimenez (2006) of developed plane channel flow at Re = 87000 (based on channel height and bulk velocity) performed on 1.8×10^{10} grid points using 2048 processors of a supercomputer during an entire half year. Hence, at present and also in the foreseeable future, DNS is not a method for practical calculations. However, it is a very useful tool for studying the details of turbulence at lower Reynolds numbers as the complete information on all details of the turbulent motion can be extracted from such simulations. As an example, results from the channel-flow simulation of Hoyas and Jimenez (2006) are presented in Figure 1.8 showing by way of the spanwise vorticity the structure of turbulence across the channel. As second example, Figure 1.9 shows a near-wall hairpin vortex structure extracted from a

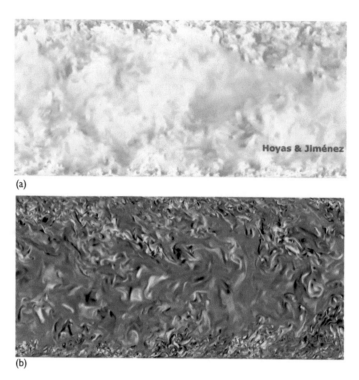

Figure 1.8 Structures from DNS of closed channel flow at Re$_\tau$ = 2000 (Re = 87.000) visualized by wall-normal velocity component (a), spanwise vorticity (b); the top and bottom boundaries are no-slip smooth walls, from Hoyas and Jimenez (2006).

channel-flow DNS at a lower Reynolds number, demonstrating that the DNS method allows a detailed study of the development of such structures.

Because until a few decades ago one could not even think of solving the Navier Stokes equations for turbulent flow, statistical methods were developed in which the turbulent fluctuations are averaged out and only equations governing mean-flow quantities are solved. These are the Reynolds-averaged equations and these methods are hence called RANS methods (for Reynolds-Averaged Navier-Stokes). Splitting up the instantaneous flow quantities into mean and fluctuating values and then averaging the non-linear original Navier Stokes equation leads to the appearance of correlations between velocity fluctuations, which act like stresses on the mean flow and are called turbulent or Reynolds stresses. They represent transport of mean momentum by the turbulent fluctuations. Similarly, averaged scalar transport equations contain correlations representing the transport of heat or mass by the turbulent fluctuations. Hence, the correlations appearing in the averaged equations express the effect of turbulence on the mean quantities. These terms are unknown and must be described by a model before the mean-flow equations can be solved. Models of this kind, which have to account for the effect of the entire spectrum of turbulent motions in a calculation are called statistical turbulence models or RANS models. It is methods using such models that are presently used in most practical calculations in hydraulics. The computing

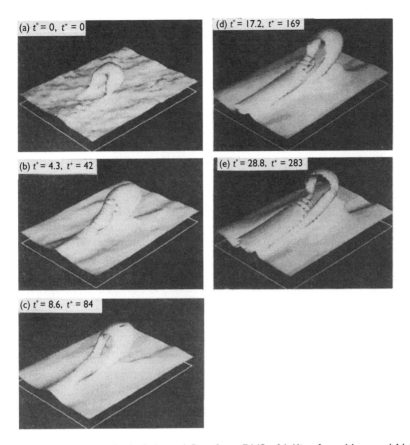

(a) $t^* = 0$, $t^+ = 0$

(b) $t^* = 4.3$, $t^+ = 42$

(c) $t^* = 8.6$, $t^+ = 84$

(d) $t^* = 17.2$, $t^+ = 169$

(e) $t^* = 28.8$, $t^+ = 283$

Figure 1.9 Hairpin vortices at bed of channel flow from DNS of J. Kim, from Nezu and Nakagawa (1993), see also Kim (1987).

effort for solving the RANS equations yielding the mean quantities is of course much less than that required by a DNS resolving the fluctuating turbulent motion at all scales. Hence, on computers available today RANS calculations can be performed for realistic situations occurring in practice, even for complex geometries and for larger flow regimes and river stretches, especially when 2D depth-averaged models are employed. A wide variety of RANS turbulence models has been proposed and developed and an extensive literature exists on the subject on such models. A book by Rodi (1993) specializes on turbulence models for use in hydraulics. The book is somewhat dated (appeared first in 1979) but covers the main turbulence models still in use today. More recent books not geared on hydraulics but more general engineering applications have been published by Durbin and Pettersson-Reif (2001), Launder and Sandham (2002), Wilcox (2006), Hanjalic and Launder (2011). Turbulence models of various complexity are in use ranging from the simple mixing-length model to Reynolds-stress models employing differential transport equations for the individual Reynolds stresses. Most models used in practice employ the eddy-viscosity/diffusivity concept and estimate the eddy viscosity by relating it to the mean-flow quantities via simple algebraic relations (mixing-length model) or model transport equations for the characteristic velocity and length or time scales of turbulence (e.g. k-ε model).

RANS methods are economical, but they suffer from limited generality and have difficulties in coping adequately with many of the complex flow phenomena often found in hydraulics and environmental flow situations, especially when large-scale structures play a dominant role for the transport of momentum, heat and mass or when details of the flow, such as unsteady processes like vortex shedding or bimodal flow behaviour and unsteady forces on structures or bed elements are important and need to be resolved. RANS methods can be used in unsteady calculations (URANS), but basically they can only cope with the unsteadiness of the mean flow, e.g. when the boundary conditions are time-dependent such as in tidal channel flow. They are suitable only when there is a clear scale separation, i.e. when the time-scale of the mean-flow unsteadiness is clearly larger than the time-scale of the turbulent fluctuations. The latter cannot be resolved in such calculations. A more complete discussion on the issue will be given in Chapter 7 on Hybrid RANS-LES methods.

A method more suited and more powerful for solving problems involving the above mentioned complex phenomena is the Large-Eddy-Simulation (LES) technique. This method, which is the subject of this book, is between DNS and RANS. Like DNS it also solves the 3D time-dependent flow equations, but only for the larger-scale motions in the spectrum that can be resolved on a given numerical grid (see Figure 1.1). The motions with scales smaller than the grid size are filtered or locally averaged out and their effect must be accounted for by a Sub-Grid-Scale (SGS model). In contrast to a RANS model, which must account for the effect of the entire spectrum of the turbulent motions, a SGS model must account only for the high wave-number part of the spectrum with small-scale motions. The effect of these is mainly dissipative and in some methods (implicit LES) is achieved alternatively by using a numerical scheme which introduces some numerical dissipation (see Chapter 5 on ILES). As LES solve the 3D time-dependent Navier-Stokes and continuity equations, they are still computationally rather expensive, often because long running times are necessary to obtain reliable statistics. Away from walls, the larger turbulent eddies containing most of the energy and contributing most to the momentum, heat and mass transfer are virtually independent of Re, so that here LES does not have a Reynolds number problem as does DNS. However, near walls the length scale of turbulence decreases with increasing Re so that the number of grid points required to resolve adequately the near-wall zone increases approximately with Re^2. So, again such wall-resolving LES require so much computing effort that their application is not possible at the high Reynolds numbers often occurring in practice. Hence, special near-wall modeling is necessary and various near-wall treatments are described in Chapters 6 and 7.

1.4 SCOPE AND OUTLINE OF THE BOOK

In Computational Fluid Dynamics generally, including commercial CFD codes, LES is used more and more for calculating turbulent flows, replacing increasingly RANS, especially when complex features involving large-scale structures are important. Also in hydraulics, LES offers great potential, especially for near-field problems with complex phenomena, and can be seen as the method of the future for such problems. Hence, hydraulic engineers require knowledge about this powerful method. A number of books on and reviews of LES are available (see examples below), but these are not geared on solving hydraulic engineering problems and are mostly too theoretical for the practicing engineer. The present book aims to provide an introduction to the LES method for hydraulic and environmental engineers which is less theoretical and mathematically demanding and hence, hopefully, easy to understand and which covers special features of flows in water bodies and summarizes the experience gained with LES for calculating such flows. Readers interested in more theoretical background and wider coverage of the method and its components are referred to the following books and review articles, which are however not geared on hydraulics: Galperin and Orszag (1993/2010), Piomelli (1999), Sagaut (2006), Lesieur et al (2005), Berselli et al (2006), Kassinos et al. (2007), Grinstein et al (2007), Benocci et al. (2008) and in German language Fröhlich (2006) and Breuer (2005).

The present book introduces first the basic concept of LES (in Chapter 2) and then the various subcomponents of the method and the models entering in these. Of course not all the models and approaches proposed in the literature can be covered in the book – only those are included that are most widely used and have shown continued success, especially in hydraulics applications.

In Chapter 3, the most common subgrid-scale models are presented and discussed, while Chapter 4 provides information on numerical methods for solving the governing equations, concentrating on those features that are special in the LES context. Chapter 5 covers briefly Implicit LES (ILES) methods which do not use explicitly a subgrid-scale model but rely on numerical dissipation for damping the fluctuations at the smallest scales. In Chapter 6 methods for providing boundary conditions are presented, with special emphasis on free-surface and wall boundaries, including conditions for rough walls, and inflow boundaries. The latter pose a much more severe problem than in RANS calculations as realistic time-dependent fluctuations must generally be provided at the inflow boundary. In Chapter 7 Hybrid LES-RANS methods are covered which are particularly important for solving practical problems where at high Reynolds numbers or for large flow domains not the entire computation domain can or need be simulated by LES but part of the domain (e.g. near walls or the far field) is treated by a much less expensive RANS method. Chapter 8 concludes the methods part by presenting the most useful methods for educing and visualizing the resolved eddies and in particular the coherent structures calculated by an LES. Finally, in the extensive Chapter 9 application examples are presented for a variety of hydraulic flow situations ranging from relatively simple straight open channel flow to complex flows around structures, over beds with various roughnesses and gravity currents, thereby demonstrating the ability and the potential of the LES method. It should be noted at the end that the book deals only with genuinely 3D LES and does not cover 2D depth-average LES methods and calculations.

Chapter 2

Basic methodology of LES

2.1 NAVIER-STOKES EQUATIONS AND REYNOLDS AVERAGING (RANS)

The starting point of all simulation methods for turbulent flows, whether Direct Numerical Simulations (DNS), Large Eddy Simulations (LES) or RANS methods, are the Navier-Stokes equations, together with a corresponding equation for scalar quantities such as temperature or species concentration. For incompressible flows these equations expressing the conservation laws for mass, momentum, thermal energy/species concentration read in tensor notation[1]

Mass conservation: continuity equation

$$\frac{\partial u_i}{\partial x_i} = 0 \tag{2.1}$$

Momentum conservation: Navier-Stokes equations

$$\frac{\partial u_i}{\partial t} + \frac{\partial u_i u_j}{\partial x_j} = -\frac{1}{\rho_r}\frac{\partial p}{\partial x_i} + \nu\frac{\partial^2 u_i}{\partial x_j \partial x_j} + g\frac{\rho - \rho_r}{\rho_r} \tag{2.2}$$

Thermal energy/species concentration conservation:

$$\frac{\partial \phi}{\partial t} + \frac{\partial u_i \phi}{\partial x_i} = \Gamma\frac{\partial^2 \phi}{\partial x_i \partial x_i} + S_\phi \tag{2.3}$$

where u_i is the instantaneous velocity component in the direction x_i, p is the instantaneous static pressure, and ϕ is a scalar quantity which may stand for either temperature T or species concentration C. S_ϕ is a volumetric source/sink term expressing, for example, heat generation due to chemical or biological reactions or the settling of suspended sediment. ν and Γ are the molecular (kinematic) viscosity and diffusivity (of ϕ) respectively. Use has been made in the above equations of the Boussinesq approximation so that the

---

1 A short introduction to tensor notation is given in Appendix A.

influence of variable density appears only in the buoyancy term, which is the last term on the right hand side of Equation (2.2) involving the reference density ρ_r and the gravitational acceleration g_i in direction x_i. Together with an equation of state relating the local density ρ to the local values of T and C, Equations (2.1) to (2.3) form a closed set and are exact equations describing all details of the turbulent motion, including all fluctuations. The Direct Numerical Simulation (DNS) method solves these equations with a suitable numerical technique, introducing no model. As discussed in the Introduction (Chapter 1), such calculations are not feasible in the foreseeable future for practical flows usually having high Reynolds numbers as the computing effort for resolving all scales including the small-scale dissipative motion would be excessive.

Hence, the RANS method has become the main, and until recently the only tool for solving practical turbulent flow problems. In this method, the turbulent fluctuations are not resolved by the calculation but averaged out by a time filter applied to the fluctuating motion. In order to make clear the difference to the spatial filtering used in LES, the basics of Reynolds averaging are briefly introduced first, restricting the discussion to the velocity field. Time-averaged/filtered mean-flow quantities are introduced and solved for and these are defined as follows:

$$<u_i> = \bar{u}_i^{RANS} = \frac{1}{T}\int_0^T u_i dt, \quad <p> = \bar{p}^{RANS} = \frac{1}{T}\int_0^T p\, dt \tag{2.4}$$

The integration/averaging time T should be much larger than the time scale of the turbulent fluctuations but smaller than the time scale of the mean motion if this is unsteady (the issue of this time-scale separation is discussed in Chapter 7). The instantaneous quantities are split up into mean values and fluctuations around the mean, e.g. for u_i:

$$u_i\left(x_i, t\right) = \bar{u}_i^{RANS}\left(x_i, t\right) + u'_{iRANS}\left(x_i, t\right)$$

Introduction of this into the continuity and Navier-Stokes Equations (2.1) and (2.2) and averaging according to the procedure in Equation (2.4) yields the Reynolds Averaged Navier-Stokes (RANS) equations governing the mean-flow quantities $\bar{u}_i^{RANS}$ and $\bar{p}^{RANS}$ calculated by the RANS method. These equations are formally identical to Equations (2.9)–(2.11) given later. When performed on the non-linear term $\partial u_i u_j/\partial x_j$ in the Navier-Stokes equations, the splitting and averaging procedure leads to the introduction of correlations between fluctuating velocities $<u'_{iRANS}u'_{jRANS}>$. These act like stresses and represent the Reynolds stresses[2] $\tau_{ij}^{RANS} = -<u'_{iRANS}u'_{jRANS}>$ expressing the entire effect of turbulence on the mean motion and require modelling, i.e. a RANS model. Such a model accounts for all scales, including the effect of the large, energy-containing ones that depend strongly on the boundary conditions and hence the flow situation considered. For this reason, as was explained in the Introduction, it has been found impossible to develop a general model which can cope with all situations of practical interest.

---

2 Strictly, for dimensional reasons, $-\rho<u'_{iRANS}u'_{jRANS}>$ are the stresses.

2.2   THE IDEA OF LES

As was explained already in the Introduction, turbulence is a multiscale phenomenon with a wide spectrum of scales of the fluid motion. Large and small scale
motions have quite different features, and these differences are summarized in
Table 2.1. The idea of LES is to calculate explicitly the motion of large scales
(or eddies) by solving the governing 3D time-dependent equations and to model
the motions of the small scales. This avoids the problem of the RANS method of
having to model also the large-scale, energetic and boundary-condition dependent
motion and at the same time the problem of DNS of having to resolve the small-
scale dissipative motion. As Table 2.1 indicates, the latter, which represents only
a small part of the spectrum, is more universal and hence easier to model than the
large-scale motion, and it has been found that quite simple models are often sufficient. The fact that the small-scale motion does not have to be resolved removes
the restriction of the method to low Reynolds number situations, which is inherent
in DNS.

The concept of LES is illustrated in Figure 2.1 vis à vis the energy flux in turbulent
flows. The large scales extract energy from the mean flow and transfer it to smaller
scales in the energy cascade, and at the small scales, which are modelled in an LES,
the kinetic energy is withdrawn by the mechanism of dissipation.

The first step in realizing the LES idea is to separate the turbulent motion
into large scales or eddies to be resolved and small scales to be modelled. Several
approaches for this separation will be introduced below. Ideally the separation
should occur in a spectral region where only energy transfer takes place (i.e. no
energy input from the mean motion, no dissipation), that is in the inertial subrange
(see spectrum in Figure 2.1) so that clearly the energetic, boundary-condition
dependent eddies are resolved and only the dissipative motion needs to be modelled. This is not always feasible, but at least the resolved eddies should contain
most of the energy (say 80–90%). The scale separation is in practice often dictated
by the grid fineness that can be afforded. LES can hence also be considered as
a method that resolves as much of the motion as possible on a given/affordable
grid.

Table 2.1 Differences between large eddies and small scale turbulence (adapted
from Schumann 1993).

Large eddies	Small scale turbulence
• produced by mean flow	• produced by large eddies
• depend on geometry and boundaries	• universal
• ordered	• random
• require deterministic description	• can be modelled statistically
• inhomogenous	• homogenous
• anisotropic	• isotropic
• long-living and energetic	• Short living and non-energetic
• diffusive	• dissipative
⇒ difficult to model	⇒ easier to model
⇒ universal model impossible	⇒ universal model possible

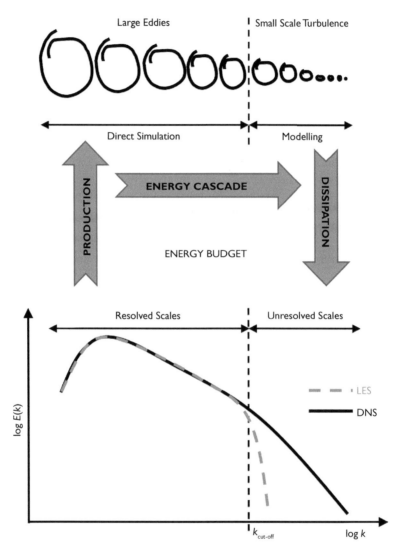

Figure 2.1 Concept of Large Eddy Simulation in relation to energy flux and energy spectrum (top part after Breuer, 2002).

2.3 SPATIAL FILTERING/AVERAGING AND RESULTING EQUATIONS

As opposed to the RANS approach involving time averaging or filtering, in LES the small-scale motion is removed by spatial averaging or filtering. Local quantities f are then split into resolved quantities $\bar{f}$ and deviations f' from these (see Figures 2.4, 2.5):

$$f = \bar{f} + f' \tag{2.5}$$

In practice, the removal of the small-scale motion and hence the averaging is performed mostly by the numerical grid as on a given grid only motions with scales larger than the mesh size can be resolved; the others fall through the mesh, and the quantity that is calculated is an average over the control volume formed by the grid (see Figure 2.2). This is the method of Schumann (1975), which introduces directly the discretizised volume-balance equations related to the numerical solution. A method independent of this is filtering. Since this method, which was first proposed by Leonard (1974), is more general and also conceptually clearer, it will be introduced first. The resolved quantity $\bar{f}$ is in the filter method defined by

$$\bar{f}(\mathbf{r},t) = \int_D G(\mathbf{r},\mathbf{r}',\Delta)f(\mathbf{r}',t)dV' \tag{2.6}$$

Here $\mathbf{r}$ is the location where $\bar{f}$ is to be determined and $\mathbf{r}'$ is the location where f is considered in the spatial integration; the integration is performed over the entire flow domain D, and G is a compactly supported (i.e. $\bar{f}$ assumes large values when $\mathbf{r}' \to \mathbf{r}$) filter function with filter width Δ, normalized so that

$$\int_D G(\mathbf{r},\mathbf{r}',\Delta)dV' = 1 \tag{2.7}$$

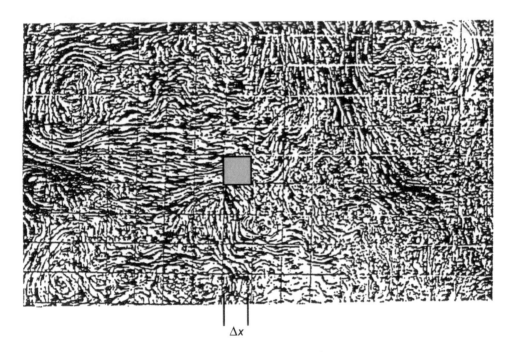

Δx

Figure 2.2 Resolution of cell-averaged quantities.

For one spatial direction (2.6) reads:

$$\overline{f}(x_i,t) = \int_{D_i} G(x_i, x_i', t) f(x_i', t) dx_i' \qquad (2.8)$$

The most commonly used filter functions are shown in Figure 2.3. These are the top-hat (or box) filter $G = \frac{1}{\Delta}$ for $|x - x_i| \le \Delta/2$ corresponding to averaging in this region, otherwise $G = 0$, the Gauss filter and the cut-off filter. The latter corresponds to cutting off all fluctuations beyond a certain wave number $k_{cut\text{-}off}$ in the spectrum while top-hat and Gaussian filters lead to a spectral distribution as shown in Figure 2.1 by the line designated LES. The effect of applying top-hat filters of different widths Δ to the spatial distribution of a quantity f is shown in Figure 2.4. The larger the filter width Δ, the larger is the part of the small-scale fluctuations filtered out and

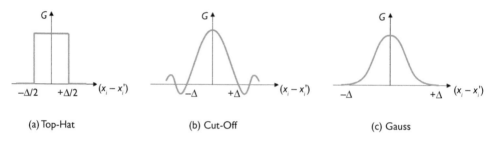

| (a) Top-Hat | (b) Cut-Off | (c) Gauss |

Figure 2.3 Filter functions G commonly used in LES.

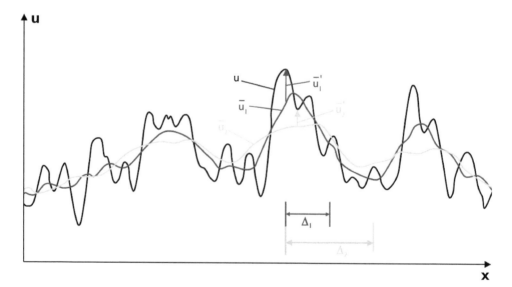

Figure 2.4 Filtered functions $\overline{u}$ obtained from $u(x)$ by applying top-hat filters of different filter width Δ (adapted from Fröhlich 2006).

the smoother the variation of $\overline{f}$. It should be noted that, as opposed to Schumann's volume-balance approach, filtering leads to a continuous $\overline{f}$-function. Also, in general, applying the filter twice smoothes further the f-distribution (i.e. $\overline{\overline{f}} \neq \overline{f}$) and also $\overline{f'} \neq 0$. These and other features of filtering are discussed in detail in the relevant literature (e.g. Pope 2000, Sagaut 2006, Lesieur et al. 2005, Breuer 2002, Fröhlich 2006).

Applying the filter operation to the Equations (2.1)–(2.3) leads to the following filtered equations governing the resolved quantities in LES:

Continuity equation: $\dfrac{\partial \overline{u_i}}{\partial x_i} = 0$ (2.9)

Navier-Stokes equations: $\dfrac{\partial \overline{u_i}}{\partial t} + \dfrac{\partial \overline{u_i}\,\overline{u_j}}{\partial x_j} = -\dfrac{1}{\rho_r}\dfrac{\partial \overline{p}}{\partial x_i} + \dfrac{\partial}{\partial x_j}\left(\nu\dfrac{\partial \overline{u_i}}{\partial x_j}\right) - \dfrac{\partial \tau_{ij}^{SGS}}{\partial x_j} + g_i\dfrac{\overline{\rho}-\rho_r}{\rho_r}$ (2.10)

Scalar transport equation: $\dfrac{\partial \overline{\phi}}{\partial t} + \dfrac{\partial \overline{u_i}\,\overline{\phi}}{\partial x_i} = \dfrac{\partial}{\partial x_i}\left(\Gamma\dfrac{\partial \overline{\phi}}{\partial x_i}\right) - \dfrac{\partial q_i^{SGS}}{\partial x_i} + \overline{S}_\phi$ (2.11)

The non-linear term $u_i u_j$ in the Navier-Stokes equations (2.2) leads originally to the filtered quantity $\overline{u_i u_j}$ in the convection term. When this is expressed as convection of the resolved quantities $\overline{u_i}\,\overline{u_j}$, the difference

$$\tau_{ij}^{SGS} = \overline{u_i u_j} - \overline{u_i}\,\overline{u_j}$$ (2.12)

represents the effect of the unresolved fluctuations on the resolved motion, acting like stresses which are therefore called subgrid-scale stresses[3]. These stresses need to be modelled by a subgrid-scale model (see Chapter 3). They are analogous to the Reynolds stresses τ_{ij}^{RANS} in the RANS approach, but while the latter represent the effect of the entire turbulent fluctuations on the mean motion, τ_{ij}^{SGS} only accounts for the effect of the small-scale motion. Similarly, in the filtered scalar transport equation (2.11) a term

$$q_i^{SGS} = \overline{u_i \phi} - \overline{u_i}\,\overline{\phi}$$ (2.13)

appears which is the subgrid-scale turbulent flux and represents the effect of the unresolved small-scale motion on the resolved scalar field.

When writing according to Equation (2.5) the unfiltered velocities u_i and u_j appearing in the terms on the RHS of (2.12) as sum of resolvable velocities ($\overline{u_i}$) and unresolved fluctuations (u_i'), τ_{ij}^{SGS} can be split up into 3 terms, namely the Leonard stresses describing the interaction of fluctuations of the larger-scale resolvable field, cross terms standing for the interaction of resolvable and unresolvable fluctuations, and a fine-scale term representing the interaction of unresolvable fluctuations, $\overline{u_i' u_j'}$.

3 In the filter approach τ_{ij}^{SGS} should strictly be called subfilter-scale stresses, but subgrid-scale stresses is the generally adopted phrase.

As it is the entire stress term τ_{ij}^{SGS} that appears in the filtered equation (2.10) and needs to be modelled, only models for this are presented in Chapter 3 and the individual terms are not dealt with further. Discussion on these can be found in the standard LES literature.

2.4 IMPLICIT FILTERING AND SCHUMANN'S APPROACH

The introduction of an explicit filter as described above resulting in a continuous smoothed function $\bar{f}$ (see Figure 2.4) allows separating the scale division from the discretization and hence from numerical effects as the filter width Δ can be larger than the mesh size. Hence, with sufficiently fine grids and small time steps, one can get for a particular SGS model a solution that is independent of these numerical parameters. However, this approach is only followed in fundamental studies and is not what is done or attempted in practice – there rather the grid size is the filter width and the filter function G does not appear explicitly in LES codes. The approach in practice is to solve Equations (2.9)–(2.11) numerically on a certain grid, usually chosen as fine as one can afford. Thereby the cell-average values are the quantities solved for, as introduced already in Figure 2.2 and illustrated for a 1D distribution in Figure 2.5. This figure also shows that there is a close relation to top-hat filtering, as when the filter width Δ is equal to the mesh size Δx_i then the cell-average value is equal to the filtered value at the grid point. However, the average values are constant over the control volumes around discretisized points and there is no continuous distribution. The scale separation now depends on the discretization of the flow field, and so does the SGS

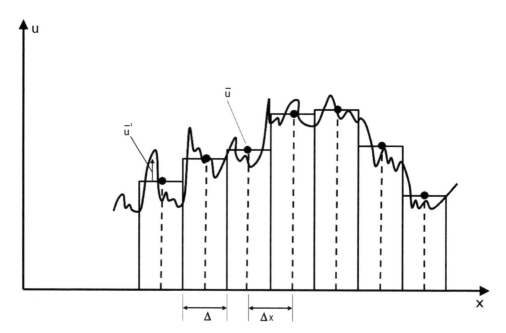

Figure 2.5 Discontinuous distribution of cell-averaged velocities $\bar{u}$ and identical values at grid points from top-hat filtering when filter width Δ equals the mesh size Δx (after Breuer, 2002).

model to be used. Hence, the solution obtained for the resolved quantities cannot be grid/time-step independent.

This implicit filtering approach is closely related to the volume-balance method proposed by Schumann (1975). The derivation of the method starts from a given finite-volume mesh and the original equations (2.1)–(2.3) are integrated analytically over individual finite volume cells yielding discrete balance equations for determining the volume-averaged value $\bar{f}$ as resolved quantity shown in Figure 2.5. Restricting further discussion to the Navier-Stokes equations (2.2), applying the Gauss theorem in the integration process to the convection terms $u_i u_j$ yields area integrals over the various faces of the control volume, $\overline{^j u_i u_j}$, where the superscript j indicates the direction normal to the face considered. Since the individual velocities vary over the faces due to the small-scale fluctuations, the face-average convection flux $\overline{^j u_i u_j}$ is not generally equal to the convection flux calculated from the resolved velocities, $\overline{^j u_i}\,\overline{^j u_j}$, except when all fluctuations are resolved as in a DNS. Hence, the difference appears in the balance equation and acts as stress on the control-volume faces. These stresses, τ_{ij}^{SGS}, are due to the unresolved subgrid-scale motion and need to be modelled, as in the filter approach. The face-averaged velocities $\overline{^j u_i}$ and $\overline{^j u_j}$ need to be related to the cell-averaged values $\bar{u}_i$ resulting from the solution of the discrete balance equations. For this an interpolation needs to be introduced, and this is discussed in Chapter 4 on Numerical Methods. The resulting discrete equations are in fact the same as those resulting from the filter approach when a top-hat filter is used, the filter width is equal to the grid size and the same discretization approximations for solving numerically the equations are employed. Hence, in practice the Schumann approach and filtering are only symbolically different, and the fact that explicit filtering is not really applied in practice is the reason why the filter approach is not dealt with in greater detail in this book and the interested reader is referred to the relevant LES literature.

2.5 RELATION OF LES TO DNS AND RANS

When the filter width or grid size is small enough so that the fluctuations of all scales are resolved, the subgrid-stresses τ_{ij}^{SGS} go to zero and the original Navier-Stokes equations (2.1)–(2.2) are solved so that the simulation is a DNS. This occurs automatically because subgrid-scale models yield $\tau_{ij}^{SGS} \to 0$ when the filter width/grid size approaches zero. In the other extreme, i.e. the RANS method, all fluctuations are (time-)filtered/averaged out and only mean-flow quantities are solved for; and when the mean flow is unsteady, the method is called URANS and the time-filtering/averaging removes only the turbulent fluctuations, but not the lower-frequency unsteadiness of the mean flow. In any case, the effect of the entire turbulence, i.e. its fluctuations at all scales, is represented by the Reynolds stresses, τ_{ij}^{RANS}, and needs to be accounted for by a model. In URANS, the equations for solving the mean-flow quantities are the same as the LES equations (2.9)–(2.11) for solving the resolved quantities, and both contain a turbulent stress term τ_{ij} that needs to be modelled. The difference lies in the model for this stress – while τ_{ij}^{SGS} requires only a model for the effect of the unresolved small-scale motion, τ_{ij}^{RANS} requires a model for the entire spectrum of the fluctuations. This difference, and the relation between LES and URANS is discussed further in Chapter 7 on Hybrid methods where both LES and RANS methods are combined in one way or another. One important aspect should however be pointed out here: a RANS model involves velocity and length scales of the turbulence that are characteristic of the energetic, larger-scale motion contributing most to the Reynolds stresses, and the results of a RANS/URANS calculation should be independent of the numerical solution of the equations, including the spatial and temporal discretization. On the other hand, in LES the SGS model depends on the scale separation, i.e. the filter width and in practice generally the grid size, which determines the length scale of the small-scale motion to be modelled by the SGS model; and when the grid size determines the scale separation, the solution is grid dependent and as the discretization is refined approaches the DNS solution.

Chapter 3

Subgrid-Scale (SGS) models

3.1 ROLE AND DESIRED QUALITIES OF AN SGS-MODEL

As shown in the last chapter, spatial filtering/averaging introduces Subgrid-Scale (SGS) stresses τ_{ij}^{SGS} and scalar fluxes q_i^{SGS} in the equations for the resolved quantities used in LES, representing the effects of the unresolved, small-scale turbulence on the resolved/ filtered motion. One way to account for this effect is through an explicit SGS model for τ_{ij}^{SGS}. As the effect is mainly dissipative, i.e. to withdraw energy from the resolved motion, another approach is to account for it through numerical dissipation introduced by the solution procedure, and this is described briefly in Chapter 5 on Implicit Large-Eddy Simulation (ILES) methods. However, in hydraulic-flow calculations, the most common method is still the use of an explicit SGS model, and in this chapter the most widely used SGS models are introduced.

The primary goal of an SGS-model is to dissipate the correct amount of energy from the directly calculated large-scale flow and to allow for a physically realistic exchange of energy between the resolved scales. A successful SGS model should then yield accurate statistics of the energy-containing scales of motion that are resolved by the simulation. The most important interactions to be modelled by the SGS model are those between the largest unresolved (subgrid) scales and the smallest resolved scales. In a well-resolved LES, the boundary between the resolved (larger) and unresolved (smaller) scales is situated within the inertial subrange (indicated by the thin dashed line in the spectrum in Figure 2.1). The effect of the SGS model on the spectral distribution of energy is visible in the vicinity of the cut-off wave number. Beyond this wave number, i.e. at smaller scales, the SGS model in a LES causes much faster energy dissipation than in the DNS. Figure 3.1 demonstrates the effect of filter width and hence cut-off wave number on the resulting spectrum in a flow. In the left spectrum a fairly large filter width is used and hence a small cut-off wave number. Thus, the amount of unresolved scales is relatively large so that more SGS modeling effort is required, especially when the unresolved scales represent energy-containing, anisotropic eddies. This approach is commonly referred to as Very-Large-Eddy Simulation (VLES). On the other hand, when a considerably smaller filter width is chosen so that the cut-off wave number is at the other end of the inertial subrange (the right spectrum in Fig. 3.1), more of the motion is resolved and less modelling effort is required so that quite simple SGS models can be employed.

In a well-resolved Large-Eddy Simulation energy conservation is achieved, hence one important requirement of a SGS-model is to provide physically correct dissipation,

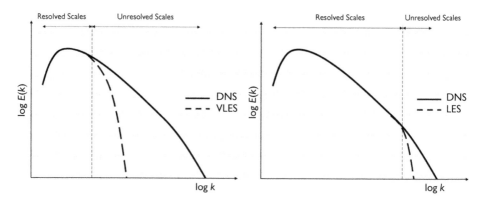

Figure 3.1 Effect of filter width on the energy spectra of LES vs. a DNS spectrum. LES spectrum with a large filter width (left) and LES spectrum with a small filter width (right).

which is the only way the turbulent kinetic energy is removed from the resolved scales. A too large dissipation will artificially increase the diffusive fluxes leading to excessive damping of the resolved scales (the left spectrum in Fig. 3.2), decreasing the accuracy of the simulation. A too small SGS-model dissipation will result in a pile up of energy around the cut-off wavenumber (the right spectrum in Fig. 3.2). This results in inaccurate flow statistics or in the numerical solution becoming unstable.

Generally, the subgrid-scale stress tensor τ_{ij}^{SGS} is split into an isotropic and an anisotropic component as:

$$\tau_{ij}^{SGS} = \underbrace{\tau_{ij}}_{anisotropic} + \underbrace{\frac{1}{3}\tau_{kk}^{SGS}\delta_{ij}}_{isotropic} \tag{3.1}$$

The isotropic part of the SGS stress tensor contains the sum of the SGS normal stresses τ_{kk}^{SGS} which is twice the kinetic energy k^{SGS} of the SGS fluctuations and acts like a pressure. This component is therefore usually added to the filtered pressure term, which leads to a new pressure variable:

$$P = \overline{p} + \frac{1}{3}\tau_{kk}^{SGS} \tag{3.2}$$

This separation of the isotropic stress from the anisotropic part is convenient when employing SGS models relating τ_{ij} to the gradients of the resolved velocity via an eddy viscosity. This is in direct analogy to the use of such a relation in RANS models. The main modelling effort is then shifted to the determination of the eddy viscosity ν_t. In the following, the most commonly used eddy viscosity models are introduced in some detail, such as the purely algebraic Smagorinsky model and its dynamic variant, the WALE model, and a model employing a transport equation for the subgrid-scale fluctuating kinetic energy k^{SGS}. Only a brief overview is given of further models not based on the eddy-viscosity concept, as these are not so widely used.

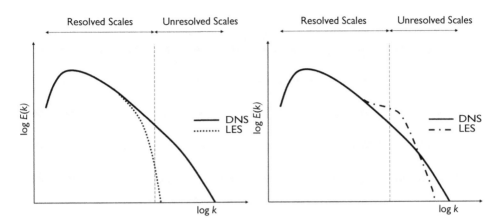

Figure 3.2 Effect of SGS energy dissipation on the energy spectra of LES vs. a DNS spectrum. LES spectrum with a highly dissipative SGS model (left) and LES spectrum with a low dissipative SGS model (right).

3.2 SMAGORINSKY MODEL

The most popular eddy-viscosity SGS model is the Smagorinsky model (Smagorinsky, 1963). In analogy to the viscous stress in laminar flows, the anisotropic stress tensor τ_{ij} is approximated by relating it to the resolved rate of strain, $\bar{S}_{ij}$, which involves velocity gradients, via an artificial eddy (or turbulent) viscosity v_t as

$$\tau_{ij} = -2v_t \bar{S}_{ij} \quad \text{in which} \quad \bar{S}_{ij} = \frac{1}{2}\left(\frac{\partial \bar{u}_i}{\partial x_j} + \frac{\partial \bar{u}_j}{\partial x_i}\right) \tag{3.3}$$

It must be emphasized that the eddy viscosity is not a fluid property but characterizes the unresolved sub-grid-scale fluctuations and depends on the resolved velocity field, $\bar{u}_i$. Due to the decomposition of the SGS stress into an isotropic part and an anisotropic part in Equation (3.1), the latter can be combined with the viscous stress term in Equation (2.10) and can be treated together in a numerical procedure.

From dimensional analysis it follows that

$$v_t \propto \ell q \tag{3.4}$$

where ℓ and q are respectively characteristic length and velocity scales of the sub-grid-scale motion.

The selection of the characteristic length-scale ℓ is in LES much simpler and more straightforward than in RANS modelling. In LES, the largest scales of the unresolved turbulence, which interact most actively with the resolved motion, are of the size of the filter width Δ. Hence, the characteristic length-scale in the Smagorinsky model is chosen as:

$$\ell = C_S \Delta \tag{3.5}$$

in which C_S is the Smagorinsky constant, an empirical parameter whose value can be obtained from theoretical considerations or *a-priori* and *a-posteriori* tests. The determination of the characteristic velocity scale can be done in analogy to Prandtl's mixing length theory, with the advantage that in LES the length scale is known already. This leads to:

$$q = \ell \cdot \left| \overline{S}_{ij} \right| = C_S \Delta \cdot \left| \overline{S}_{ij} \right| \quad \text{with} \quad \left| \overline{S}_{ij} \right| = \sqrt{2 \overline{S}_{ij} \overline{S}_{ij}} \tag{3.6}$$

which yields the eddy viscosity as:

$$\nu_t = \ell \cdot q = \ell^2 \left| \overline{S}_{ij} \right| = (C_S \Delta)^2 \cdot \left| \overline{S}_{ij} \right| \tag{3.7}$$

This relation for the eddy viscosity can also be derived by assuming local equilibrium between production and dissipation of sub-grid-scale kinetic energy k^{SGS}, i.e. by equating the last two terms in Equation (3.27).

The model has one adjustable parameter, the Smagorinsky constant C_S, which is assumed constant in the original formulation of the model. For isotropic turbulence, Lilly (1992) predicted $C_S = 0.165$ based on assuming local equilibrium in the inertial subrange. For shear flows (e.g., channel flows) optimum values were found to be $C_S = 0.065 - 0.1$, which can yield a six times difference in the values of ν_t predicted by the Smagorinsky model. An important shortcoming of the Smagorinsky model is that ν_t as predicted by Equation (3.7) does not reduce to zero in the viscous sublayer, in which turbulent fluctuations should be damped as impermeable surfaces, e.g. walls, are approached. This is due to the fact that large velocity gradients prevail in the boundary layer and result in high values of the rate of strain $\left| \overline{S} \right|$. Hence, the turbulent eddy viscosity ν_t needs to be damped near impermeable surfaces by a damping function. The most popular of such $f(z^+)$ was proposed by van Driest (1956) in the mixing length RANS model. It is used in the Smagorinsky model to reduce the length scale ℓ:

$$\ell = C_S \Delta \cdot f(z^+) \quad \text{with} \quad f(z^+) = 1 - e^{-z^+/A^+}, \tag{3.8}$$

where $z^+ = \frac{z u_*}{\nu}$ [1] is the normal distance to the wall z in wall units and A^+ is an empirical constant ($A^+ = 25$). The van Driest damping function is easy to implement but does not always produce accurate results. This is partly due to the fact that with the original van Driest damping (Eq. 3.8) all modelled SGS stresses scale linearly with the distance from the wall i.e. $\tau_{ij} \propto (z^+)$, while the most relevant SGS components of the SGS stress tensor, i.e. the shear stress τ_{13}; τ_{23} should scale with the cubed distance from the wall, i.e. $\tau_{ij} \propto (z^+)^3$ (Hinze, 1975). Hence, Piomelli et al. (1989) proposed a slightly modified version of the van Driest damping function:

$$f(z^+) = \sqrt{(1 - e^{-z^+/A^+})^3}, \tag{3.9}$$

---

1 often u_τ is used instead of u_*.

which provides the correct behaviour for the most relevant components of the SGS stress tensor and is currently more in use than Equation (3.8).

A second shortcoming of the Smagorinsky model is that it cannot be used to calculate laminar-turbulent transition since the eddy viscosity is only zero if velocity gradients are absent, so that $\nu_t \geq 0$ while in laminar flow ν_t should be zero. It should be noted, however, that transition is not of practical relevance in hydraulic flows. The fact that in turbulent flows $\nu_t \geq 0$ implies that the kinetic energy is transferred only from large resolved scales to small scales, i.e. the Smagornisky model is strictly dissipative and does not allow for a backscatter of energy from small unresolved scales to large scales.

Furthermore, in complex highly three-dimensional flows (e.g. flow around bluff bodies) the optimum value of C_S varies locally within the flow domain and is basically impossible to determine a priori. The Smagorinsky constant needs to be further modified to account for rotation and stratification effects.

Despite the above-mentioned shortcomings of the Smagorinsky model, it has been widely applied to many turbulent flow studies in hydraulic engineering. This is due to its simplicity in terms of calculating the SGS stresses via an algebraic relation and in terms of implementation into a code, for which the total viscosity term can be introduced as:

$$\nu_{total} = \nu + \nu_t \tag{3.10}$$

The Smagorinsky model furthermore has only one adjustable parameter, whose effect on the statistics of the flow vanishes when decreasing the filter width.

Several modified eddy viscosity models were proposed to alleviate the deficiencies of the classical Smagorinsky model. They include the use of a modified velocity scale for the largest unresolved scales which also results in a model that does not need wall damping (WALE model), the use of a dynamic procedure to calculate the model coefficient (dynamic Samgorinsky model) and solving a transport equation for the SGS kinetic energy k^{SGS} which provides the velocity scale of the largest unresolved scales (one equation SGS models). These three approaches will be discussed in the following subsections.

3.3 IMPROVED VERSIONS OF EDDY VISCOSITY MODELS

3.3.1 Dynamic procedure

The main idea of the dynamic procedure is to calculate the model parameters of a base model (e.g. the constant C_S when applied to the Smagorinsky model) by using information available from the smallest resolved scales (Germano et al., 1991). To achieve this, Germano et al. (1991) suggested to introduce a second filter, a test filter, with width $\tilde{\Delta}$ larger than the original filter Δ, and to employ the same model for calculating the sub-grid-scale stresses τ_{ij} and the stresses T_{ij} resulting from the sub-test-filter motions (see Figure 3.3). In analogy to (2.12) defining τ_{ij}, the stresses T_{ij} read

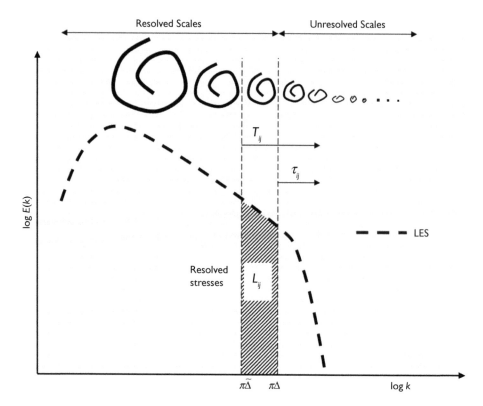

Figure 3.3 Typical energy spectrum of an LES that employs a dynamic procedure SGS model.

$$T_{ij} = \widetilde{\overline{u_i u_j}} - \widetilde{\overline{u}}_i \widetilde{\overline{u}}_j \qquad (3.11)$$

where the tilde represents the second (test) filter operation. When the correlation $\overline{u_i u_j}$ in (3.11) is decomposed into the contribution $\overline{u}_i \overline{u}_j$ resolved on the grid Δ and the stress τ_{ij} and the second filter is applied, there follows the relation known as Germano's identity:

$$T_{ij} = -L_{ij} + \tilde{\tau}_{ij} \qquad (3.12)$$

where

$$L_{ij} = -\widetilde{\overline{u}_i \overline{u}_j} + \widetilde{\overline{u}}_i \widetilde{\overline{u}}_j \qquad (3.13)$$

representing the part of the sub-test-scale stresses that are resolved between $\tilde{\Delta}$ and Δ (see Figure 3.3), i.e. the smallest resolved scales. This term can be calculated explicitly from the resolved velocities.

For determining the value of the constant C_S in the Smagorinsky model from the smallest resolved velocities, this model (Equations (3.3) and (3.7)) is now applied to represent the anisotropic part of both T_{ij} and τ_{ij} in (3.12):

$$\tau_{ij} = -2(C_S \Delta)^2 \left|\bar{S}_{ij}\right| \bar{S}_{ij} \tag{3.14}$$

$$T_{ij} = -2(C_S \tilde{\Delta})^2 \left|\tilde{\bar{S}}_{ij}\right| \tilde{\bar{S}}_{ij} \tag{3.15}$$

in which the double-filtered strain rate tensor is defined as:

$$\tilde{\bar{S}}_{ij} = \frac{1}{2}\left(\frac{\partial \tilde{\bar{u}}_j}{\partial x_i} + \frac{\partial \tilde{\bar{u}}_j}{\partial x_i}\right) \quad \text{and} \quad \left|\tilde{\bar{S}}_{ij}\right| = \sqrt{2\tilde{\bar{S}}_{ij}\tilde{\bar{S}}_{ij}} \tag{3.16}$$

Inserting T_{ij} from (3.15) and τ_{ij} from (3.14) with second filter applied into (3.12) yields:

$$L_{ij} = 2(C_S\Delta)^2\left[\frac{\tilde{\Delta}^2}{\Delta^2}\left|\tilde{\bar{S}}_{ij}\right|\tilde{\bar{S}}_{ij} - \overline{\left|\bar{S}_{ij}\right|\bar{S}_{ij}}\right] = 2(C_S\Delta)^2\,M_{ij} \tag{3.17}$$

in which

$$M_{ij} = \frac{\tilde{\Delta}^2}{\Delta^2}\left|\tilde{\bar{S}}_{ij}\right|\tilde{\bar{S}}_{ij} - \overline{\left|\bar{S}_{ij}\right|\bar{S}_{ij}} \tag{3.18}$$

From the filtered velocity field, i.e. $\bar{u}_i$, a test-filtered velocity field can be calculated explicitly by applying a filter procedure with a larger filter width. Filtered and test-filtered stress tensors, i.e. $\bar{S}_{ij}$ and $\tilde{\bar{S}}_{ij}$, as well as the test-filtered products of the stress tensors or velocities, i.e. $\overline{\left|\bar{S}_{ij}\right|\bar{S}_{ij}}$ and $\widetilde{u_i u_j}$ can then also be calculated explicitly. Hence L_{ij} (using Equation 3.11) and M_{ij} (using Equation 3.18) can be computed to obtain the constant C_S. However, since Equation (3.17) is a tensor equation, it contains six independent equations to determine one parameter C_S and can only be satisfied approximately or in some average sense. It should also be mentioned here that the ratio of the filter width $\tilde{\Delta}/\Delta$ appearing in (3.18) is usually chosen as 2.

For closed-channel flow Germano et al. (1991) suggested to average over horizontal planes parallel to the upper and lower wall. Lilly (1992) improved the original averaging procedure by suggesting a least-square procedure to be applied to minimize the error $e_{ij} = L_{ij} - 2(C_S\Delta)^2 M_{ij}$ in Equation (3.16). This allows obtaining the value of C_S that best satisfies the over-determined system of Equation (3.17). For this particular case, the norm of the error is:

$$E^2 = e_{ij}e_{ij} = L_{ij}^2 - 2(C_S\Delta)^2 L_{ij}M_{ij} + 4(C_S\Delta)^4 M_{ij}M_{ij} \tag{3.19}$$

The norm of the error is minimum when $dE^2/dC = 0$, in which $C = C_S^2$. This allows estimating C from

$$C\Delta^2 = \frac{1}{2}\frac{L_{ij}M_{ij}}{M_{ij}M_{ij}} \tag{3.20}$$

Calculating the parameter $C\Delta^2$ at each time step based on the resolved flow field is called the dynamic procedure. An advantage of the dynamic-procedure version of the Smagorinsky model (DSM) is that it predicts the behaviour of ν_t near solid surfaces correctly without the need for empirical damping functions. Furthermore, the DSM predicts zero values of ν_t in laminar flow, and does not require special corrections to account for rotational and stratification effects.

As Equation (3.20) predicts $C\Delta^2$ rather than C, the DSM is independent of the definition of the turbulence length scale (e.g. $\Delta = (\Delta_1\Delta_2\Delta_3)^{1/3}$ or $\Delta = (\Delta_1^2 + \Delta_2^2 + \Delta_3^2)^{1/2}$, where Δ_i are the mesh sizes in the individual coordinate directions x_i). This is advantageous if mesh cells are strongly anisotropic.

Regularization

Local and instantaneous negative values of C (or $C\Delta^2$) and hence eddy viscosities are possible, which, in theory, corresponds to the backscatter effect, i.e. an energy transfer from small unresolved scales to large scales. However, negative eddy viscosities tend to destabilize the numerical procedure and should be avoided. Furthermore, strong variations of C in space and time can also lead to numerical instabilities. One partial remedy is a procedure called clipping, which limits the computed eddy viscosity to values greater than zero i.e. $\nu_t > 0$, or the total viscosity to be greater than zero, i.e. $\nu + \nu_t > 0$. Far more common than the clipping procedure is the application of some sort of averaging procedure, either in space or in time, depending on the flow problem. One possibility is to average the nominator and denominator in Equation (3.20) over one or more homogeneous directions in the flow (if any exist). By applying the spatial averaging operator $\{\ \}$, Equation (3.20) can be rewritten as:

$$C\Delta^2 = \frac{1}{2}\frac{\{L_{ij}M_{ij}\}}{\{M_{ij}M_{ij}\}} \tag{3.21}$$

In complex three-dimensional flow, i.e. flows in which there is no homogeneous direction, the spatial distribution of C as predicted by the dynamic procedure can be vastly heterogeneous across small regions. This high spatial variability can also generate numerical instabilities, so that instead of averaging over homogeneous flow directions, a local averaging and/or filtering of the predicted field of C is applied. As a (negative) consequence, the result depends on the volume chosen for averaging or on the stencil of the filtering operator. Alternatively, Akselvoll and Moin (1993) applied an averaging procedure in time for fully inhomogeneous flows. They have chosen a special form of time averaging which acts like a lowpass filter:

$$C_{filtered}^{n+1} = (1 - fi)\, C^n + fi\, C^{n+1} \tag{3.22}$$

here C^n is the value of C from the previous time step. Breuer and Rodi (1994) applied this methodology to the channel flow in a bend. They chose values of fi of the order 10^{-3} so that all high frequency oscillations could be damped out and only the low frequency variations remained.

An alternative averaging approach for applying the dynamic procedure in flows with complex configurations was proposed by Meneveau et al. (1996). These authors suggested using Lagrangian averaging, i.e. the averages in Equation (3.21) are taken over a fluid particle path line (backwards in time).

By using Lagrangian averaging in the estimation of the model coefficient, the number of points at which the model predicts negative values and the spatial variability in the values of the dynamic coefficient is reduced. As a result, the robustness of the numerical simulation is improved. The *Lagrangian Dynamic Model* can also be combined with other models such as the scaled dependent dynamic model (Porte-Agel et al., 2000) and dynamic mixed models (Zang et al., 1993). Sarghini et al. (1999) provide a detailed comparison of the performance of the dynamic mixed model with and without Lagrangian averaging with the dynamic Smagorinsky model and with the constant coefficient Smagorinsky model for 2D channel flows and 3D boundary layers. The use of Lagrangian averaging was found to improve the accuracy of the results compared to implementations in which local or plane averaging was used to estimate the nominator and denominators in the expressions for the model coefficients.

Further methods exist for improving the robustness of the dynamic procedure for the Smagorinsky model, especially for inhomogeneous flows. One such approach is the dynamic localization model of Ghosal et al. (1995) in which an integral equation is solved to determine the model coefficient. Several simpler alternatives of the dynamic localization model exist that reduce the computational overhead related to solving exactly the integral equation (e.g. Piomelli and Liu, 1995).

3.3.2 WALE model

The Wall-Adapting Local Eddy-viscosity (WALE) model (Nicoud and Ducros, 1999) has found increasing interest recently as it is a relatively simple eddy-viscosity model that can account for wall effects without employing wall-damping functions. Nicoud and Ducros (1999) proposed to use information from the resolved velocity-gradient tensor $g_{ij} = \partial \bar{u}_i / \partial x_j$ to calculate the eddy viscosity. The WALE model uses the traceless symmetric part of the square of g_{ij} to calculate the eddy viscosity as:

$$v_t = (C_w \Delta)^2 \frac{\left| (G_{ij}^a) \right|^{6/2}}{(\bar{S}_{ij} \bar{S}_{ij})^{5/2} + \left| (G_{ij}^a) \right|^{5/2}} \tag{3.23}$$

where G_{ij}^a is the traceless part of $G_{ij} = 1/2(g_{ik}g_{kj} + g_{jk}g_{ki})$. C_w is a model constant for which values in the range of 0.45–0.5 were recommended based on information extracted from simulations of isotropic homogeneous turbulence (Nicoud and Ducros, 1999). One of the advantages of the model is that it predicts correctly the behaviour of the eddy viscosity near solid surfaces i.e. $v_t = O(z^3)$. Another advantage is that the WALE model can be applied to complex geometries with either structured or unstructured grids because no explicit filtering associated with the introduction of a test filter as in the dynamic procedure is needed. Moreover, despite using a constant coefficient, the model predicts a zero value of v_t in laminar shear flow (e.g., in particular in the case of a wall bounded laminar flow) and can be used to correctly simulate flows with regions in which relaminarization or transition to turbulence occur.

3.3.3 Transport-equation SGS models

The above introduced SGS models relate the SGS stress tensor locally to the resolved velocity field, ignoring thereby non-local and history effects on the SGS stresses. The simplest way to account for such effects is to use instead of the SGS velocity scale according to Equation (3.6) the turbulent kinetic energy $k = \frac{1}{2}\tau_{kk}$ of the SGS motions as the (square of the) SGS velocity scale and to solve a transport equation for k. Hence

$$q = k^{1/2} \tag{3.24}$$

and there follows from Equation (3.4) for the eddy viscosity (Schumann, 1975):

$$v_t = C_v \Delta k^{1/2} \tag{3.25}$$

and hence for the SGS stresses

$$\tau_{ij} = -2C_v \Delta k^{1/2} \overline{S}_{ij} \tag{3.26}$$

here C_v is a model constant. In the original model of Yoshizawa (1982), the SGS kinetic energy k is determined from the following model transport equation

$$\frac{\partial k}{\partial t} + \frac{\partial}{\partial x_j}(\overline{u}_j k) = \frac{\partial}{\partial x_j}\left[(v + C_k \Delta k^{1/2})\frac{\partial k}{\partial x_j}\right] + 2C_v \Delta k^{1/2}\overline{S}_{ij}\overline{S}_{ij} - C_\varepsilon \frac{k^{3/2}}{\Delta}, \tag{3.27}$$

where C_ε and C_k are additional constants. In Yoshizawa and Horiuti (1985), the model constants were estimated as $C_v = 0.05$, $C_k = 0.1$ and $C_\varepsilon = 1.0$.

The above transport-equation SGS model is analogous to eddy-viscosity based transport-equation RANS models. However, in the latter the turbulent length scale appearing in the eddy-viscosity relation and in the dissipation term of the k-equation has to be estimated empirically or determined from an additional equation (ε-equation or ω-equation in k-ε or k-ω models, respectively), while in the SGS model the relevant length scale is the specified filter width Δ.

As was mentioned already, neglecting the history and transport terms (terms on the left hand side and 1st term on the right hand side) and equilibrating the remaining production and dissipation terms yields directly the Smagorinsky model according to Equation (3.7). This is again in analogy to RANS models where the assumption of local equilibrium between production and dissipation of turbulent kinetic energy leads to the zero-equation mixing-length model.

In order to allow a variation and adjustment of the model coefficient, the dynamic procedure can be employed to estimate C_v (Menon et al., 1996; Davidson, 1998) as:

$$C_v = \frac{1}{2} \frac{L_{ij} M'_{ij}}{M'_{ij} M'_{ij}} \tag{3.28}$$

with L_{ij} being the Leonard stress (Eq. 3.11) and M'_{ij} defined as:

$$M'_{ij} = -\tilde{\Delta} K^{\frac{1}{2}} \tilde{\tilde{S}}_{ij} + \Delta k^{\frac{1}{2}} \widetilde{S_{ij}} \tag{3.29}$$

where K is the kinetic energy calculated using the test filtered velocities.

In most one-equation SGS models the eddy viscosity in the momentum equations and in the k-equation are identical so that $C_v = C_k$. In analogy to the dynamic Smagorinsky model, the constant C_v can attain negative values. While negative values of C_v are not critical in the k-equation, the value of C_v in the momentum equations needs extra treatment to ensure numerical stability. Krajnovic and Davidson (2002) suggest a regularization procedure, similar to the procedure used for the dynamic Smagorinsky model.

In principal, one-equation SGS models have several advantages over the standard Smagorinsky model or the dynamic Smagorinsky model, respectively. An important advantage is the fact that the one-equation model allows for backscatter without destabilization of the numerical solution procedure, because the k-equation requires an SGS energy balance. Ghosal et al. (1995) show that this approach is stable and provides always positive values for the SGS energy. One-equation models furthermore provide the correct asymptotic behaviour near solid walls and allow for simulation of transitional flows (Ghosal et al., 1995). Another advantage is that, in theory, no averaging procedure is required when employing a dynamic procedure to determine C_v and C_ε. On the other hand, in one-equation models it is necessary to solve one extra transport equation, which makes the model computationally more expensive than the dynamic Smagorinsky SGS model. Thus far, one-equation models have been employed primarily in the meteorological community, e.g. Moeng (1984), and to predict flows with combustion (Kim and Menon, 2000), which is because other effects (e.g. buoyancy, large-scale SGS roughness, chemical reactions) play a role and for which solving a transport equation offers advantages.

3.4   SGS MODELS NOT BASED ON THE EDDY VISCOSITY CONCEPT

3.4.1   Scale-Similarity Model

The Scale-Similarity Model (SSM) of Bardina et al. (1980) does not use the eddy-viscosity concept. Rather, the idea is to assume that the smallest resolved scales are similar to the largest unresolved scales and to use that information to obtain an expression for the SGS stresses τ_{ij}. This is justified as the most important interactions are those between the largest unresolved (subgrid) scales and the smallest resolved scales. To obtain an expression for τ_{ij} one has first to define these scales. By definition, the unresolved velocity u'_i (at length scale smaller than Δ) can be written as $u'_i = u_i - \bar{u}_i$, where u_i is the (unfiltered) velocity component and $\bar{u}_i$ the filtered velocity. The velocity of the largest unresolved scales is defined as $\overline{u_i'}$, yielding with the above decomposition $\overline{u_i'} = \overline{u_i} - \overline{\overline{u_i}}$. Bardina et al. (1980) suggested to also filter the resolved field $\overline{u_i}$ yielding $\overline{\overline{u_i}}$ in order to obtain the smallest resolved scales $\hat{u}_i$ by subtracting $\overline{\overline{u_i}}$ from the resolved velocity, i.e. $\hat{u}_i = \overline{u_i} - \overline{\overline{u_i}}$. This shows that $\overline{u_i'} = \hat{u}_i$ which is justified by assuming that the corresponding scales have similar structure near the cutoff wavenumber. The SGS stresses can now be assumed to be the same as the ones due to the resolved field $\overline{u_i}$, i.e.

$$\tau_{ij} = \overline{u_i u_j} - \overline{u_i}\,\overline{u_j} \approx C_B\left(\overline{\overline{u_i}\,\overline{u_j}} - \overline{\overline{u_i}}\,\overline{\overline{u_j}}\right), \tag{3.30}$$

where C_B is the Bardina constant, for which usually a value of $C_B = 1.0$ is assumed. Compared to predictions obtained with the Smagorinsky eddy-viscosity model, the SSM model showed an improvement in predicting SGS stresses as it can account for backscatter in a physical way. However, the SSM model does not dissipate enough energy from the large scales and is, in most applications, combined with a dissipative model such as the Smagorinsky model, yielding a mixed model.

3.4.2   Dynamic Mixed Model

Two versions of the Dynamic Mixed Model (DMM) are worth mentioning. In both models, the SSM model is combined with the Smagorinsky model. The DMM as proposed by Zang et al. (1993), assumes the Bardina constant to be unity and combines the SSM model with the original Smagorinsky model in the following way:

$$\tau_{ij} = \overline{\overline{u_i}\,\overline{u_j}} - \overline{\overline{u_i}}\,\overline{\overline{u_j}} - 2C_S^2\Delta^2\left|\overline{S}_{ij}\right|\overline{S}_{ij} \tag{3.31}$$

In the DMM proposed by Salvetti and Banerjee (1995), both coefficients i.e. the Bardina coefficient C_B and the Smagorinsky constant C_S are determined dynamically. In complex turbulent flows, the use of this variant of the DMM has shown that the spatial and temporal variations of C_S are reduced substantially compared to those predicted by the dynamic Smagorinsky model. Even without averaging, the numerical model is much more stable. However, numerical instabilities can still develop if at some locations the backscatter is large and occurs over large periods of time.

3.4.3   Approximate Deconvolution Models (ADM) and Sub-Filter Scale Models (SFS)

The idea of Approximate Deconvolution Models (ADM) or Sub-Filter Scale (SFS) models, respectively, is to use the information from the filtered (resolved) velocity field to reconstruct, the subgrid (unresolved) and unfiltered quantities. This can, in theory, be done by an inverse filtering, also called deconvolution. However, inverse filtering requires information from the unresolved scales so that the reconstruction can only be approximated. In the Approximate Deconvolution Model (ADM), Stolz and Adams (1999) use a truncated series expansion of the inverse filter to approximate the unfiltered velocity field. This approximation of the unfiltered variables is then used to compute the nonlinear terms in the filtered Navier–Stokes equations, which avoids the need to compute additional subgrid scale terms. Chow et al. (2005) argue that for accurate simulation of high Reynolds number boundary layer flows the resolvable Subfilter-Scale (SFS) stresses should not be neglected. They proposed to use explicit filtering and reconstruction of the velocity field using Taylor series expansion to calculate the resolvable SFS-stresses. For the SGS stresses a dynamic eddy-viscosity model is employed. By combining reconstruction and eddy-viscosity models (e.g. the Smagorinsky model) higher order versions of Bardina's mixed model are obtained.

The advantage of the ADM or SFS models is that they contain no parameters and can account for backscatter in a physical way. In *a-priori* and *a-posteriori* tests, e.g. turbulent channel flow (Stolz et al., 2001, Gullbrand and Chow, 2003) comparisons between filtered DNS stresses and the ADM or SFS model showed that the stresses predicted by the two models gave improved agreement with DNS over the standard Smagorinsky SGS-model predictions. Another interesting approach is the one of Stoltz et al. (2005) who proposed a high-pass filtered eddy viscosity (Smagorinsky) model in which the variable model coefficients are determined by high-pass filtering of the resolved variables. The model does not need wall-damping functions to correctly predict the viscous sublayer of wall bounded turbulent flows nor a dynamic estimation of the model coefficient. Its performance is similar to that of the dynamic Smagorinsky model (see discussion in Stolz et al., 2007).

3.5 SGS MODELS FOR THE SCALAR TRANSPORT EQUATION

In calculations with scalar transport using the LES method, a model for the subgrid scalar fluxes is needed for the solution of the filtered scalar equation. The subgrid-scale scalar fluxes q_i^{SGS} appearing in the filtered transport equations (2.11) for a conserved or non-conserved scalar $\bar{\phi}$ (e.g., contaminant, concentration of suspended sediment, temperature) can be modelled in direct analogy to eddy-viscosity models through a gradient-diffusion SGS model:

$$q_i^{SGS} = \overline{u_i \phi} - \bar{u}_i \bar{\phi} = \Gamma_t \frac{\partial \bar{\phi}}{\partial x_i}, \tag{3.32}$$

where Γ_t is the SGS eddy diffusivity. The simplest approach for calculating Γ_t is to assume proportionality between eddy diffusivity and eddy viscosity through the turbulent Schmidt number, Sc_t (or in case of temperature the turbulent Prandtl number, Pr_t) so that:

$$\Gamma_t = \frac{v_t}{Sc_t} \text{ or } \Gamma_t = \frac{v_t}{Pr_t} \tag{3.33}$$

Typical values for Pr_t, or Sc_t are in the range 0.3–0.7 (Deardorff, 1974, Moeng, 1984, Andren et al., 1994).

The use of constant and prescribed turbulent Schmidt or Prandtl numbers can be avoided by employing a dynamic procedure for calculating the SGS turbulent scalar fluxes in analogy to the one for calculating the SGS-eddy viscosity (Moin et al., 1991). The eddy diffusivity Γ_t can be calculated with the same algebraic formulation as the one used for the eddy viscosity (Equation 3.7) as:

$$\Gamma_t = C_\Gamma \Delta^2 \cdot |\bar{S}_{ij}| \tag{3.34}$$

The coefficient C_Γ, the analogue of the dynamic Smagorinsky constant in the dynamic procedure, is calculated using the resolved velocity and scalar fields:

$$C_\Gamma \Delta^2 = \frac{1}{2} \frac{L_i'' M_i''}{M_i'' M_i''} \tag{3.35}$$

in which

$$L_i'' = \widetilde{\overline{u_i} \overline{\phi}} - \widetilde{\overline{u}}_i \widetilde{\overline{\phi}} \tag{3.36}$$

is the analogue to the Leonard stress (Eq. 3.13) and

$$M_i'' = \frac{\tilde{\Delta}^2}{\Delta^2} |\tilde{\bar{S}}_{ij}| \frac{\partial \tilde{\bar{\phi}}}{\partial x_i} - \widetilde{|\bar{S}_{ij}| \frac{\partial \bar{\phi}}{\partial x_i}} \tag{3.37}$$

is the scalar analogue to M_{ij} (given by Equation (3.18)). As in the dynamic Smagorinsky model, the test-filtered scalar field can be computed from the resolved (filtered) scalar field. Again an averaging procedure is required to calculate the coefficient C_Γ. Recent simulations of flow and mass transfer processes over cavities in which both eddy viscosity and eddy diffusivity were computed using the dynamic procedure provided accurate predictions (Chang et al., 2006, Constantinescu et al., 2009).

Chapter 4

Numerical methods

4.1 INTRODUCTION

The governing partial differential equations for LES that were introduced in previous chapters need to be solved with an adequate numerical method, which is comprised of various components. These components are (a) approximation of the derivatives through algebraic operators (b) discretization of the physical domain with a computational grid that consists of a finite number of cells, points or elements, respectively, at which the continuous functions of the variables are represented and (c) solution of the resulting system of algebraic equations for discrete instants in time. In this chapter, some fundamentals of numerical methods are introduced but only aspects of special relevance to LES are discussed in more detail. For an in-depth treatment of the diverse discretization schemes, grid generation techniques and solution procedures of the algebraic equations, the interested reader is referred to standard CFD textbooks (e.g. Wendt, 1996; Ferziger and Peric, 2002, Versteeg and Malalasekera, 2007, Hirsch, 2007).

The goal of LES is to simulate three-dimensional, unsteady, turbulent flows and hence methods that were developed for laminar flows or that have been successful in the context of RANS may not be adequate for LES. Moreover, in LES the unsteadiness of the flow that is comprised of eddies of different size, frequency, energy content and longevity, respectively needs to be reproduced correctly. This puts a high demand on the numerical method, which needs to meet certain requirements in terms of its accuracy, however, as will be discussed below, a high order of the method alone does not guarantee a correct reproduction of the flow physics.

In general, the accuracy of a large-eddy simulation depends on:

i   the selected discretization scheme to approximate the derivatives in space and time and to some extent its formal order of accuracy
ii   grid spacing and time step; that is how closely the discrete system approximates the continuum.
iii   the capabilities of the subgrid-scale model used
iv   the adequacy of the boundary conditions
v   the solution method for solving the incompressible flow equations

The order of accuracy of the discretization scheme in space and time (i) and the grid spacing and time step (ii) are closely related because the truncation error of the numerical approximation of the derivatives (see 4.2) is generally smaller the finer the grid and the smaller the time step. It should be pointed out already here and will be shown later in the chapter that high-order numerical schemes, in spite of their high formal accuracy, may not be suitable in LES because they may not provide an accurate description of the large-scale eddies and their transport and decay into smaller ones. Furthermore, in most LES the filter width and hence the size of the smallest resolved eddies is specified by the mesh size, and it is therefore the motion of these eddies that is most negatively affected by the truncation error of the numerical scheme. On the other hand, exactly these scales provide information that is used in the subgrid-scale model. Hence, in LES there is a strong interaction between the subgrid-scale model employed and the numerical scheme as well as the grid resolution. This is particularly evident in the method of Implicit Large Eddy Simulation (ILES), which gains popularity and is covered briefly in Chapter 5. The basis of ILES is that, for instance, upwind (biased) schemes produce a certain amount of numerical dissipation, which results in removal of energy mainly from the smaller resolved scales, thereby substituting and avoiding the use of an explicit subgrid-scale model.

In general, the finer the grid, the smaller is the portion of the spectrum that requires modelling and hence the impact of both the subgrid-scale model and the numerical scheme on the solution process. Therefore, the capabilities of the subgrid-scale model (iii) become particularly important when the LES is carried out on coarser grids. This is of relevance in LES of hydraulic interest as in practice the number of grid points and hence the grid spacing that can be achieved is often determined by the available computer hardware.

An important aspect of numerical methods for LES is stability because in LES the computed variable distributions are non-smooth and progressively more heterogeneous as the Reynolds number increases. Hence, certain stability conditions need to be obeyed, which generally results in an upper limit of the time step. Due to the strong interconnection between temporal and spatial scales in LES, the time step should, however, not only be set to achieve numerical stability but should also be compatible to the grid spacing from a physical point of view, as will be explained below in 4.2.3.

In most LES, and in particular those of practical hydraulic interest, the flow is physically bounded (e.g. free surface, rough walls, inflow boundary), for which adequate boundary conditions are needed (iv). These are often referred to as supergrid models, as they introduce approximations, and inadequate or unphysical boundary conditions can introduce errors to the solution that can easily exceed the errors due to truncation or due to subgrid-scale modeling by several orders of magnitude.

The incompressibility of the working fluid implies that the flow must be divergence free at every instant in time. As a consequence, momentum and continuity equations need to be coupled, resulting in a solution method of the system of governing equations that is prone to inaccuracies (v). In LES, the solution method chosen requires careful consideration with respect to maintaining the conservation properties. In most LES, the above mentioned coupling is achieved through a Poisson-type equation for the pressure, which is solved implicitly using a suitable matrix solution procedure. Matrix solvers, however, are not covered in this book, because these are general methods for which the interested reader is referred to standard CFD textbooks (e.g. Ferziger and Peric, 2002, Hirsch, 2007).

In subchapter 4.2, the most often used methods in LES for discretizing the governing equations in space and time are introduced. In subchapter 4.3, several of the numerical methods are discussed in terms of accuracy and their suitability to be employed for LES. Subchapter 4.4. points out sources of errors of numerical methods and how they can deteriorate the numerical solution. Subchapter 4.5 introduces basic concepts of the solution of the governing equations for incompressible flows and how LES is affected. Finally, in 4.6 a number of possibilities on how to discretize complex LES domains are presented. Obviously, this chapter does not (and is not meant to) cover all details of numerical methods in CFD, but is rather tailored to LES in order to safeguard high quality simulations for proficient and new LES users.

4.2   DISCRETIZATION METHODS

The main numerical methods used in LES are the following:

i   Finite Difference Method (FDM)
ii   Finite Volume Method (FEM)
iii   Finite Element Method (FEM)
iv   Spectral Method (SM)

By far the two most common methods used in LES are finite-difference and finite-volume methods. A few LES have been carried out with the finite element method (e.g. LES of flow around tube bundles by Rollet-Miet et al., 1999), but finite-difference and finite-volume methods are computationally more efficient so that the FEM has not found much attention for LES. Spectral methods are the most accurate ones and offer the fastest solvers, however their numerical properties require the solution of the Navier Stokes equations in a domain with at least one homogeneous direction, cyclic boundary conditions in this direction and equidistant grids. Spectral methods are very popular for DNS and LES of decaying isotropic turbulence (e.g. Menon et al., 1996) in which the SM is used in all three spatial directions, and for channel flow (e.g. Moser et al., 1999) in which the SM is used in two directions (streamwise, spanwise) and another method, e.g. FD, in the wall-normal direction. The spectral method is very attractive for fundamental studies of turbulent flows but is not very common in the context of LES in hydraulics due to its limitation to very simple geometries. In the following, FDM and FVM are introduced and discussed in more detail.

4.2.1   Finite Difference Method (FDM)

The basic idea behind the FDM is to replace the partial derivatives of first and second order that appear in the governing equations of LES with algebraic difference quotients using values at a finite number of discrete points in the flow domain. This then leads to a system of algebraic equations for each of the flow variables, which can be solved numerically. The finite-difference method uses Taylor series expansions or polynomial fitting to derive difference quotient expressions for the derivatives at discrete grid points, expressing them through variable values at neighboring grid points. This is explained here by reference to a one-dimensional variable distribution as shown in Figure 4.1(a). This figure provides an example of a continuous function $f(x)$, being

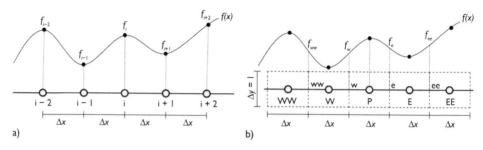

Figure 4.1 Finite difference (a) and finite volume (b) computational stencils for 1D problems.

represented as a series of discrete values f_i at discrete points (i). The value f_{i+1} at point $(i + 1)$ can be expressed in terms of a Taylor series expanded about point (i) as:

$$f_{i+1} = f_i + \left.\frac{\partial f}{\partial x}\right|_i (\Delta x)^1 + \left.\frac{\partial^2 f}{\partial x^2}\right|_i \frac{(\Delta x)^2}{2} + \left.\frac{\partial^3 f}{\partial x^3}\right|_i \frac{(\Delta x)^3}{6} + \cdots \tag{4.1}$$

Similar expansions can be made for other points $(i + 2, i - 1, i - 2,$ etc). Expression (4.1) is exact if an infinite number of terms on the right hand side is retained and/or if $\Delta x \to 0$. In a numerical method, Equation (4.1) is truncated and the accuracy of the solution depends on which terms are neglected. The more terms are neglected, the lower is generally the accuracy of the solution. This can be demonstrated by investigating the Finite Difference (FD) approximation of the first derivative of a function $f(x)$, $\frac{df}{dx}$. With (4.1) and similar Taylor series expansions for other neighboring points, the following approximations can be derived:

1st Order
$$\left.\frac{\partial f}{\partial x}\right|_i = \frac{f_{i+1} - f_i}{\Delta x} + \mathcal{T}_1 \tag{4.2a}$$

2nd Order
$$\left.\frac{\partial f}{\partial x}\right|_i = \frac{f_{i+1} - f_{i-1}}{2\Delta x} + \mathcal{T}_2 \tag{4.2b}$$

3rd Order
$$\left.\frac{\partial f}{\partial x}\right|_i = \frac{-f_{i+2} + 6f_{i+1} - 3f_i + 2f_{i-1}}{6\Delta x} + \mathcal{T}_3 \tag{4.2c}$$

4th Order
$$\left.\frac{\partial f}{\partial x}\right|_i = \frac{-f_{i+2} + 8f_{i+1} - 8f_{i-1} + f_{i-2}}{12\Delta x} + \mathcal{T}_4 \tag{4.2d}$$

The above represent the first order forward (or upwind) difference (4.2a), the third order forward-biased difference (4.2c), and the second (4.2b) and fourth (4.2d) order central-difference approximations. The truncation term, $\mathcal{T}_m$, represents the higher order terms not accounted for in the difference approximations and is the difference between the exact (Taylor) solution of the derivative and its discrete approximation.

For instance, the truncation terms of the 1st order forward difference and the 2nd order central difference are:

$$\mathcal{T}_1 = -\frac{(\Delta x)^1}{2} \frac{\partial^2 f}{\partial x^2}\bigg|_i - \frac{(\Delta x)^2}{6} \frac{\partial^3 f}{\partial x^3}\bigg|_i - \cdots = O(\Delta x^1) \tag{4.3a}$$

$$\mathcal{T}_2 = -\frac{(\Delta x)^2}{6} \frac{\partial^3 f}{\partial x^3}\bigg|_i - \frac{(\Delta x)^4}{120} \frac{\partial^5 f}{\partial x^5}\bigg|_i - \cdots = O(\Delta x^2) \tag{4.3b}$$

The neglect of the truncation terms introduces an error into the finite-difference approximation, and the rate at which this decreases as Δx decreases determines the accuracy of the approximation, respectively the order of the scheme (m). Comparing the dominant term in (4.3a) with the one in (4.3b) it is seen that, as Δx goes to zero, the second order truncation error, $\mathcal{T}_2$, approaches zero much faster than $\mathcal{T}_1$. Hence, using the same grid, the 2nd order finite-difference approximations of the first derivative are more accurate than 1st order finite-difference approximations.

The 2nd order finite-difference approximation of the second derivate, $\frac{\partial^2 f}{\partial x^2}$, is obtained by substituting Equation (4.2b) for the first derivative in Equation (4.1), neglecting terms with third and higher derivatives, and solve for the second derivative:

$$\text{2nd Order} \qquad \frac{\partial^2 f}{\partial x^2}\bigg|_i = \frac{f_{i-1} - 2f_i + f_{i+1}}{\Delta x^2} + \mathcal{T}_2 \tag{4.4}$$

This most commonly used central-difference approximation for the second derivative is second-order accurate, because, on a uniform grid, the leading term in the truncation error involves $(\Delta x)^2$. This approximation is second-order accurate also on non-uniform grids (Ferziger and Peric, 2002). Higher-order approximations (e.g. 4th order) for the second derivative can be obtained by including more neighboring points, however a fourth order scheme for the second derivative is only reasonable when convective terms are discretized using fourth order or higher approximations. Furthermore, in convection-dominated flows (as in hydraulic engineering) the gain in accuracy when using higher order approximations for the diffusion terms is minimal and does not warrant the extra computational effort.

In finite-difference schemes (as well as in any other numerical scheme) the accuracy of a simulation depends on both the grid spacing and the chosen difference quotient approximations of the first and second derivative. An important property of finite-difference approximations is that, as the grid spacing approaches zero, all terms in the truncation error approach zero as well. Finite-difference approximations that exhibit such asymptotic behaviour are called consistent, which is an important asset of any numerical solution. However, the choice of consistent and accurate approximations of the spatial derivatives is only one aspect to consider when performing LES, as will be discussed further in sections 4.3 and 4.4., which deal in more detail with numerical accuracy and numerical errors with regard to LES.

In general, finite-difference methods are easy to implement into a Navier Stokes Solver, but the finite-difference method necessitates the computational grid to be structured (see 4.6.1), which is quite a restrictive requirement when dealing with complex, three-dimensional geometries as are common in hydraulics. The restriction can be relieved by multi-block grids (see 4.6.2), by using the FDM on curvilinear coordinates, by making use of the immersed-boundary method (see 4.6.4), or by a combination of the three techniques.

4.2.2 Finite Volume Method (FVM)

In the FV method the governing differential equations are integrated over a finite number of Control Volumes (CVs) that comprise the flow domain (an example of a 2D finite-volume discretization is given in Figure 4.5). This results in a balance equation for each CV that expresses the rate of change of a quantity in the CV as the sum of its flux through the CV faces, and in the case of momentum also of pressure forces acting on the CV faces and of volume forces (such as gravity). Here, the balance equation is, for the sake of simplicity, derived for the general 1D unsteady convection-diffusion equation (without source terms), which reads:

$$\frac{\partial f}{\partial t} + u \frac{\partial f}{\partial x} = \Gamma \frac{\partial^2 f}{\partial x^2}, \tag{4.5}$$

in which u is the convective velocity and Γ is the diffusion coefficient and, again for the sake of simplicity, u and Γ are assumed constant. The most common finite-volume approach is to represent the computational domain by a suitable numerical grid and then locate the computational node at the centroid of the CV (Ferziger and Peric, 2002). This is sketched in Figure 4.1(b) for a Cartesian 1-D finite-volume domain with the CV around node P, for which integration of (4.5) between $w(est)$ and $e(ast)$ yields:

$$\int_w^e \frac{\partial f}{\partial t} dx + \int_w^e u \frac{\partial f}{\partial x} dx = \int_w^e \Gamma \frac{\partial^2 f}{\partial x^2} dx \tag{4.6}$$

resulting in

$$\frac{\partial}{\partial t} \underbrace{\left(\frac{1}{\Delta x} \int_w^e f dx \right)}_{\bar{f}} \Delta x + \underbrace{uf_e - uf_w}_{C = net\ convective\ flux} = \Gamma \underbrace{\left[\frac{df}{dx}\bigg|_e - \frac{df}{dx}\bigg|_w \right]}_{D = net\ diffusive\ flux} \tag{4.7}$$

In Equation (4.7) the transient term represents the rate of change of the average quantity $\bar{f}$ in the CV, the treatment of which will be discussed below. Convection and diffusion are expressed as surface fluxes (C and D) through the two CV faces denoted e (for east) and w (for west). This flux-balance equation leads to automatic conservation of the quantity considered, one of the advantages of the finite-volume method. As can be seen from equation (4.7), in the FVM the variable values and their gradients at the cell faces appear and have to be obtained by interpolation. The simplest interpolation is a linear one using nodal values of neighboring CVs (W and E in Fig. 4.1b). For a uniform grid, the net convective flux then yields:

$$C = u[f_e - f_w] = u\left[\left(\frac{f_E + f_P}{2}\right) - \left(\frac{f_P + f_W}{2}\right)\right] = u\Delta x\left(\frac{f_E - f_W}{2\Delta x}\right) \tag{4.8}$$

The term in parenthesis on the RHS of Equation (4.8) is exactly the same as derived by the FDM (i.e. Equation 4.2b), which is why the linear interpolation between nodes to approximate values at the cell faces in a finite-volume method is referred to as Central Differencing Scheme (CDS). For the net diffusive flux there follows for uniform grid spacing:

$$D = \Gamma\left[\frac{df}{dx}\bigg|_e - \frac{df}{dx}\bigg|_w\right] = \Gamma\left[\left(\frac{f_E - f_P}{\Delta x}\right) - \left(\frac{f_P - f_W}{\Delta x}\right)\right] = \Gamma\Delta x\left[\left(\frac{f_W - 2f_P + f_E}{\Delta x^2}\right)\right], \tag{4.9}$$

in which the gradients at the cell faces are estimated using the nodal values on either side of the face. Hence this is referred to as CDS scheme for the diffusive fluxes and corresponds to the FD analogue given in Equation (4.4).

Another popular interpolation assumes that the flow is convection-dominated so that, in the case of u being negative, $f_w = f_P$ and $f_e = f_E$. The resulting upwind-differencing scheme reads:

$$C = u[f_e - f_w] = u[f_E - f_P] = u\Delta x\left(\frac{f_E - f_P}{\Delta x}\right) \tag{4.10}$$

The term in parenthesis on the RHS of Equation (4.10) is equivalent to the first-order differencing expression derived in the context of the FDM (i.e. Equation 4.2a).

The challenge in FV methods is the interpolation from nodal values to CV surface values for which many different schemes of varying order are available. A number of FVM-interpolation schemes have been derived from the finite-difference analogues and can be found in standard CFD text books (e.g. Versteeg and Malalasekera, 2007; Hirsch, 2007) or in the large literature on the subject. In general, the advantage of the finite-volume method over other methods is that conservation is enforced formally in each CV volume and hence for the entire solution domain. The FV method can be used for complex geometries as it can be implemented for all types of grids. The main disadvantage, e.g. when compared to the FD method, is that FV schemes of accuracy higher than two are more difficult to develop in two or three dimensions.

4.2.3 Time discretization

The discretization of the time derivative in the filtered Navier Stokes equations using finite differences is very similar to the discretization in space, and an approximation analogous to the first-order expression (4.2a) can, for instance, be derived from a Taylor series as:

$$\frac{\partial f}{\partial t} = \frac{f^{n+1} - f^n}{\Delta t} + \mathcal{T}_1 \tag{4.11}$$

in which f^n is the value of f at time t_n, f^{n+1} is the yet unknown value of f at time t_{n+1} and $\mathcal{T}_1$ is the truncation error. For the sake of deriving the fundamental principles of time discretization, the convection-diffusion equation (4.5) is considered and is written as:

$$\frac{\partial f}{\partial t} = -u\frac{\partial f}{\partial x} + \Gamma\frac{\partial^2 f}{\partial x^2} = F \tag{4.12}$$

where F is the sum of the spatial derivative terms for convection and diffusion. Neglecting the truncation term τ_1 and combining Equations (4.11) and (4.12) yields:

$$\frac{f^{n+1} - f^n}{\Delta t} = F \tag{4.13}$$

in which the spatial derivatives in F are to be replaced by discrete approximations. Time-discretization schemes in which F is calculated using values at t_n are called explicit, because F can be calculated explicitly using known values only. If F is estimated using values at t_{n+1}, an implicit time-discretization scheme results because the right hand side of Equations (4.12) and (4.13) involve yet unknown values of f. In general, explicit time-discretization schemes are easier to program and are computationally more efficient than implicit schemes because the latter lead to algebraic difference equations involving several unknowns and require a matrix solver. On the other hand, explicit schemes face more restrictions on the time step for numerical stability. However, in LES the stability-restricted time step is also demanded by the physics of the flow, and hence many LES codes employ explicit time discretization schemes.

The simplest time-discretization schemes are explicit and implicit Euler methods, in which the variable f^{n+1} is calculated from:

$$\frac{f^{n+1} - f^n}{\Delta t} = F^n \tag{4.14a}$$

$$\frac{f^{n+1} - f^n}{\Delta t} = F^{n+1}, \tag{4.14b}$$

where superscripts n and $n + 1$ refer to the instant in time at which the spatial derivatives of the term F are calculated. Euler methods can be considered as the analogues of forward and backward differencing in space and are first-order accurate in time. Euler methods are called two-point methods, because values of f at two instances in time are involved. A second-order accurate two-point method can be constructed by applying the trapezoidal rule to approximate F, which yields the (semi)-implicit Crank-Nicholson method:

$$\frac{f^{n+1} - f^n}{\Delta t} = \frac{1}{2}\left[F^n + F^{n+1}\right] \tag{4.15}$$

It is relatively easy to construct methods of higher order by considering additional values of f in time, known as multi-point methods, or by using values between t_n and t_{n+1}, known as predictor-corrector methods. A second order explicit three-point method that is popular in LES (e.g. Thomas and Williams, 1994), is the second order Adams-Bashforth scheme, which reads:

$$\frac{f^{n+1} - f^n}{\Delta t} = \frac{1}{2}\left[3F^n - F^{n-1}\right] \tag{4.16}$$

Generally, multi-point methods may produce non-physical solutions or tend to be unstable if the time step is large, even when chosen within a given stability limit (Ferziger and Peric, 2002). One way of overcoming numerical instabilities is to use a safety factor on the chosen time step (see discussion of Equation (4.20) below) or by computing intermediate solutions of f between t^n and t^{n+1}, which is known as predictor-corrector or multi-stage method. Runge-Kutta (RK) methods are of this type; for instance, a second order RK method consists of the following two steps:

$$\frac{f^{*,\,n+1/2} - f^n}{\Delta t} = \frac{1}{2}F^n \tag{4.17a}$$

$$\frac{f^{n+1} - f^n}{\Delta t} = F^{*,\,n+1/2} \tag{4.17b}$$

where $f^{*,n+1/2}$ is a predicted value at $t_{n+1/2}$ that is being corrected in the second step to provide f^{n+1} using the predicted value $f^{*,n+1/2}$. The method is explicit and does only require values of previous time steps or the initial condition at the first time step, respectively. Since RK methods compute intermediate values between t^n and t^{n+1}, they are more stable than multi-point methods such as the Adams Bashforth method and hence are quite popular for LES. Higher-order Runge-Kutta methods are easy to develop, however the higher the order the more intermediate data need to be stored. Hence in LES either the basic second order RK method or "low storage variants" are common (e.g. Breuer, 1998a, Hinterberger et al., 2008).

The representation of the time derivative in the FVM according to (4.7) is straightforward. For the 1D case described by Figure 4.1(b) it is determined as:

$$\int_w^e \frac{\partial f}{\partial t}\,dx = \frac{\partial \bar{f}}{\partial t}\Delta x \approx \frac{\partial f}{\partial t}\Delta x \quad \text{with} \quad \bar{f} = \frac{1}{\Delta x}\int_w^e f\,dx \tag{4.18}$$

in which the volume integral is approximated as the product of the mean integrand and the volume. In the FV method it is assumed that the value at the node is an accurate estimate of the volume-averaged value $\bar{f}$. The time derivative in (4.18) is then treated in exactly the same way as in finite-difference methods and any explicit or implicit method can be employed.

In contrast to RANS, in LES the unsteadiness of the motion is of great importance and hence higher-order time-discretization schemes and small timesteps Δt are desirable. In theory, all time discretization methods produce stable solutions if Δt is sufficiently small. However, explicit time-discretization methods are subject to rigorous stability conditions, which are generally known as the CFL-condition (Courant–Friedrichs–Levy condition, Courant et al., 1928), and the Diffusion-number (DIF) condition:

$$CFL = \frac{|u|\Delta t}{\Delta x} < 1 \tag{4.19a}$$

$$DIF = \frac{\Gamma \Delta t}{\Delta x^2} < 0.5 \tag{4.19b}$$

While Equation (4.19a) is important when diffusion is small, Equation (4.19b) is important for diffusion-dominated flows. In LES (and DNS) Equation (4.19a) includes also an important physical constraint on the simulation, i.e. that on a given mesh the time variation (turbulent fluctuations) of the flow corresponding to the variation in space is resolved properly. What follows is that in LES also implicit time-discretization methods should obey the CFL condition and hence most LES use explicit time-discretization schemes, because there is then no need to solve large matrix systems through a matrix solver, which is time consuming. While the CFL condition applies in most regions of the flow, the DIF condition becomes important near solid boundaries, which is where viscous forces dominate. This leads to a combined stability criterion commonly used in LES (Miller, 1971):

$$\Delta t < \frac{fac}{\dfrac{|u_i|}{\Delta x_i} + \dfrac{2(\nu + \nu_t)}{\Delta x_i^2}}, \tag{4.20}$$

where the sum of molecular and SGS-viscosity $(\nu + \nu_t)$, replaces the diffusion coefficient Γ in Equation (4.19b), and fac is an additional safety factor (in LES usually $0.2 < fac < 0.8$) accounting for the non-linearity of the governing equations. Breuer (2002) suggests $fac = 0.2$ for Adams-Bashforth schemes and $fac = 0.6$ for Runge-Kutta based time-discretization schemes, which confirms the above mentioned stability issue of Adams Bashforth schemes when applied to solving the Navier-Stokes equations. Condition (4.20) can be quite restrictive in wall-resolving LES, where the grid spacing near the wall (especially in the direction normal to the wall) is very small. As a result, the DIF-required time steps (i.e due to the second part of the denominator in Equation (4.20) where ν remains finite) can be much smaller than what is needed in terms of reproducing an accurate time dependence of the fluctuations. To avoid the time step to be dominated by the diffusion-number condition, sometimes LES is carried out using an implicit time-advancement scheme such as the Crank-Nicholson scheme (e.g. Krajnovic and Davidson, 2002). An alternative to using an implicit scheme is to do time splitting, that is using an explicit scheme for convective terms and an implicit scheme for diffusive terms. This removes the necessity for very small time steps near the walls and hence makes it a very popular method in LES (e.g. Zang et al., 1993, Salvetti et al., 1997, Armenio et al., 1999, Omidyeganeh and Piomelli, 2011). For instance, for the LES of channel flow over ripples, Zedler and Street (2001) employ the second-order explicit Adams-Bashforth method for the convective terms and the implicit Crank-Nicholson method for the diffusive terms.

Finally, it should be mentioned that in LES the selection of both the time-discretization scheme and the time step should match the selected discretization schemes in space. If as a minimum a second order scheme is used to discretize the derivatives in space, due to the strong interconnection of temporal and spatial scales, a minimum second order time discretization is obligatory. This is particularly relevant in (almost all) LES approaches in which the grid spacing is equal (or proportional) to the filter width and hence the spatial discretization determines the frequency up to which the turbulent fluctuations are resolved and vice versa.

4.3   NUMERICAL ACCURACY IN LES

While in theory any numerical method and any approximation scheme can be used to solve the filtered Navier Stokes equations, in practice only few schemes are suitable for LES. A wide range of spatial and temporal scales and the interaction between them need to be captured in a physically correct manner. Resolution of the fluctuations requires small time steps and stable time discretization schemes of higher order. Equally important are suitable space-discretization schemes, in particular the discretization of the non-linear, convective term necessitates the use of low-diffusive, low-dispersive approximations (see discussion on diffusive and dispersive errors in the next section). What follows is that most discretization schemes used in the context of RANS, such as upwind schemes of up to 3rd order, which add artificial diffusion to the solution (demonstrated in the next section) should be rejected, because they influence negatively the energy cascade among scales (Ferziger, 1996) and damp out high frequency-components in the solution (Mittal and Moin, 1997). Hence, the choice of the discretization scheme for the convective terms in LES should not be based on the order of the scheme alone but also on other factors, such as SGS model and grid resolution. The impact of the order of the convection scheme and of grid refinement on the accuracy of the method can be demonstrated by considering the Fourier decomposition of the velocity field, which (in homogeneous turbulence) describes the distribution of the turbulent energy over a range of scales. A discrete Fourier representation of a periodic function $f(x_m)$ reads:

$$f(x_m) = \sum_{k=k_{min}}^{k_{max}} c_k\, e^{ikx_m} \tag{4.21}$$

in which $m_n = m\Delta x$, $m = 1, 2, \ldots M$, $k = 2\pi\lambda$ is the wave number, λ is the wavelength, c_k is the Fourier coefficient and $i = \sqrt{-1}$. Equation (4.21) expresses the sum of the contributions from different scales that are resolved on an equidistant grid with spacing Δx and between a minimum, $k_{min} = \pi/(M\Delta x)$ and a maximum, $k_{max} = \pi/\Delta x$, wavenumber. The derivative of $f(x_m)$, i.e. $\frac{df(x_m)}{dx}$, can be expressed with a discrete Fourier series as:

$$\frac{df(x_m)}{dx} = \sum_{k=k_{min}}^{k_{max}} ik c_k e^{ikx_m} \tag{4.22}$$

To evaluate the accuracy of finite-difference approximations, the discrete Fourier series of Equation (4.21) can be substituted into Equation (4.2), and by knowledge of the exact derivative given in Equation (4.22), an effective wavenumber, k_{eff}, is obtained, which is the numerical equivalent of the wavenumber k that can be effectively reproduced by the given discretization scheme. Ferziger (1996) and Breuer (2002) show that k_{eff} is real for second and fourth order central differencing approximations while k_{eff} is complex for first and third order differencing approximations, which indicates numerical dissipation of the schemes (Ferziger, 1996). Here, the ability of central differencing to reproduce effective wavenumbers and the effect of order and grid refinement shall be illustrated. Figure 4.2 (adapted from Breuer, 2002) presents calculated normalized (with k_{max}) effective wavenumbers, k^*_{eff}, using the second (CDS-2) and forth (CDS-4) order Central Differencing Schemes (Fig. 4.2a) and the effect of grid refinement on the calculation of the effective wavenumber (Fig. 4.2b). Also plotted in Figure 4.2 is the result using the exact Spectral Method (SM), solid line, for which $k^*_{eff} = k^*$. For small wavenumbers ($k^* < 0.2$), i.e. representing the low frequency part of the energy spectrum, the second order CDS (CDS-2) gives a very accurate representation of the exact spectrum and a fourth order CDS (CDS-4) extends this accurate representation to wavenumbers $k^* < 0.3$. However, on the given grid, the high frequency portion of the spectrum is not represented very well. A grid refinement, here for CDS-2, remedies the problem of inaccuracies at higher wavenumbers. Refining the grid (here by factors 2 and 4) allows for a better representation of the high frequency portion of the flow. Already a refinement by a factor of 2 provides better accuracy than the 4th order CDS on the original grid.

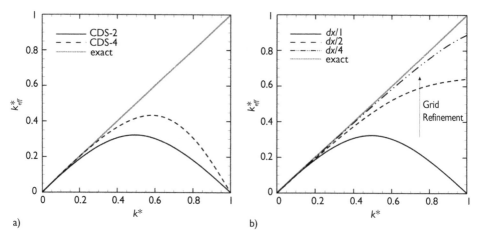

Figure 4.2 (a) Effective normalized wavenumber using second (CDS-2) and forth (CDS-4) order central differencing approximations and (b) effect of grid refinement using CDS-2 on the effective normalized wavenumber (adapted from Breuer, 2002).

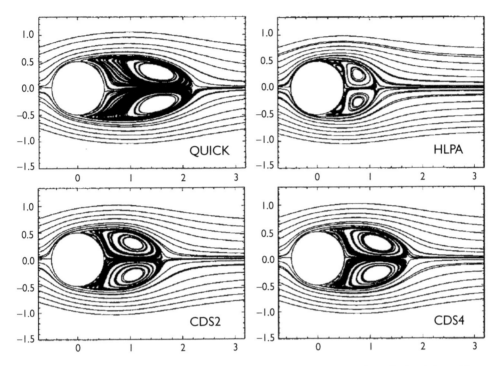

Figure 4.3 The effect of discretization scheme on the flow around a circular cylinder (adapted from Breuer, 1998b).

Though this example may only represent a relatively simple turbulent flow, it nevertheless shows that in LES the discretization scheme together with the grid resolution determine the spectrum of turbulence that is adequately resolved by the selected numerical approximation of the derivatives. On the other hand, for LES of complex flows, a-priori knowledge of how much k_{eff} deviates from k for a chosen discretization scheme or grid resolution is unattainable. Hence, numerical experiments by means of LES require some sort of validation of the chosen discretization scheme and grid resolution, for instance by using experimental or DNS data, and to carry out rigorous comparisons of measured/known turbulence statistics with the ones computed. Another validation approach is given in Ferziger (1996), who suggests quantifying the deviation of k_{eff} from k by using a measured or known energy spectrum, which can provide insight into the effects of the chosen discrete approximations and/or grid refinement on the numerical solution. On the other hand, measured or known spectra are hardly available, and deviations between measured and computed quantities may not be explained by the chosen discretization scheme alone. The reason is that in LES other (modelling) factors (SGS modelling, boundary conditions) are likely to play a bigger role than the discretization scheme. There are only few studies that have investigated the adequacy of different numerical schemes when employed in LES. Breuer (1998b) tested rigorously numerical schemes, grid resolution and subgrid-scale models for LES of the flow around a circular cylinder (Fig. 4.3). He demonstrated that central differencing schemes are superior in terms of accuracy to upwind-based schemes, regardless of the order of the scheme. In fact, the second-order central scheme (CDS2) appears to be significantly more accurate than the second-order HLPA scheme (Zhu, 1991) or the (theoretically) third-order QUICK scheme (Leonard, 1979). He also found that a fourth-order central differencing scheme (CDS4) does not provide noticeably different results than the ones obtained by CDS2. Similar conclusions were drawn by Mittal and Moin (1997), who compared turbulence statistics obtained with upwind-based and central schemes for the flow around a circular cylinder. They conclude that numerical dissipation inherent in upwind-based schemes removes substantial energy from the resolved wave-number range. This mechanism (i.e. removing energy from small scales) is, in LES, the task of the subgrid-scale model. This mechanism is being made use of in an LES approach of increasing popularity known as Implicit LES (ILES). The rationale behind ILES is that the truncation error of the discretization scheme acts as an implicit SGS model, and hence in ILES, explicit SGS models are omitted altogether. More details on ILES are provided in Chapter 5.

4.4 NUMERICAL ERRORS

As discussed above, the numerical scheme can have great influence on the accuracy and quality of LES of turbulent flows and discretization errors influence negatively the simulation results. This aspect will be discussed further by considering the 1D pure advection equation with a constant convective velocity $u > 0$:

$$\frac{\partial f}{\partial t} + u \frac{\partial f}{\partial x} = 0 \tag{4.23}$$

If the convective term is discretized using the first-order backward scheme (which is the backward analogue of Equation 4.2a) and the time derivative is discretized using the explicit Euler scheme (Equation 4.14a), the following discrete equation is obtained:

$$\frac{f_i^{n+1} - f_i^n}{\Delta t} + u \frac{f_i^n - f_{i-1}^n}{\Delta x} = 0 \tag{4.24}$$

The truncation error is $O(\Delta t, \Delta x)$ and the scheme is stable for $CFL = u\Delta t / \Delta x < 1$. Hirt (1968) proposed to replace f_i^{n+1} and f_{i-1}^n in Equation (4.24) by their Taylor series expansions around f_i^n in time and space to obtain the following so-called modified equation (also given in Hirsch, 2007):

$$\frac{\partial f}{\partial t} + u \frac{\partial f}{\partial x} = \frac{u\Delta x^1}{2}(1 - CFL)\frac{\partial^2 f}{\partial x^2} - \frac{u\Delta x^2}{6}(2CFL^2 - 3CFL + 1)\frac{\partial^3 f}{\partial x^3} + \mathcal{H} \tag{4.25}$$

where $\mathcal{H}$ represents higher order terms. The RHS of the modified equation expresses the difference of the exact solution of Equation (4.23) to the solution obtained from the discrete equation (i.e. Equation 4.24) and hence represents the numerical (truncation) error when using this approximation. The individual terms of the RHS of (4.25) can be analyzed in terms of their physical meaning. For stability purposes $CFL < 1$, hence the first term on the RHS of (4.25) is a positive, diffusion-like term that is proportional to the grid spacing Δx^1, and hence the term $\frac{u\Delta x^1}{2}(1 - CFL)$ is the numerical (or artificial) viscosity. The effect of this term on the numerical solution is that wave amplitudes are reduced, or in other words, large gradients in the solution are smoothed. The subsequent error is referred to as numerical diffusion. In the context of LES, the artificial diffusion leads to an increase in dissipation of kinetic energy, amplifying the virtue of the subgrid scale model. If the convective term of Equation (4.23) is discretized using a second-order central-differencing scheme, the term involving the second derivative on the RHS of Equation (4.25) disappears and the leading error term involves the third derivative. This impacts the numerical solution by a phase shift of the propagating waves but without altering the amplitude. This term introduces a dispersion error, which does not extract energy from the flow, but it produces numerical wiggles, which can lead to a destabilization of the entire numerical scheme.

The effect of the dissipation and dispersion errors on the numerical solution can be studied by looking at the propagation of a wave of a given amplitude (e.g., assume unit amplitude) and fixed wavenumber k. The exact solution of Equation (4.23) governing the transport of a solitary wave is given as:

$$f(x,t) = e^{ik(x-ut)}, \tag{4.26}$$

which describes the propagation of a wave of an arbitrary wavenumber with no damping or amplification of its amplitude in time. Ideally, the numerical discretization scheme should preserve the shape of the wave in the discrete solution, i.e. it should not artificially damp the amplitude of the wave or alter its propagation speed. However, due to the fact that the numerical solution is only an approximation this is not the case, and as discussed in the context of Equation (4.25), first-order approximations of the derivatives add a diffusion-like term of the form $\alpha \cdot \partial^2 f / \partial x^2$ with $\alpha > 0$ to the pure advection equation, which then takes the form of an advection-diffusion equation:

$$\frac{\partial f}{\partial t} + u\frac{\partial f}{\partial x} = \alpha\frac{\partial^2 f}{\partial x^2} \tag{4.27}$$

for which the exact solution reads:

$$f(x,t) = e^{-\alpha k^2 t} e^{ik(x-ut)} \tag{4.28}$$

The result is that the amplitude of the wave will decay with time but the speed of the wave, u, remains the same. The expression of the amplification factor in (4.28) shows that the decay is faster for large wavenumbers, k, which corresponds to small wavelengths. Hence, the presence of diffusive terms in the modified equation/truncation error damps the wave amplitude especially at high wavenumbers. The effect of diffusion is a smoothing of the solution by reducing the gradients, especially in regions where they are high which is illustrated in Figure 4.4(b).

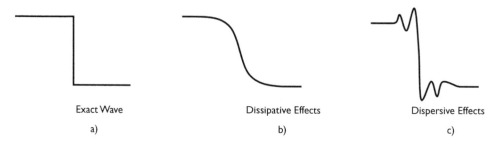

Exact Wave   Dissipative Effects   Dispersive Effects

a)   b)   c)

Figure 4.4 Sketch showing effect of dissipation and dispersion errors on the propagation of a wave. a) exact solution; b) numerical solution in which dissipation effects dominate; c) numerical solution in which dispersion effects dominate.

If the spatial derivative of Equation (4.23) is approximated with a second-order central-differencing scheme, a dispersion-like term of the form $\beta \cdot \partial^3 f / \partial x^3$ with $\beta > 0$, is added so that it becomes:

$$\frac{\partial f}{\partial t} + u \frac{\partial f}{\partial x} = -\beta \frac{\partial^3 f}{\partial x^3}, \tag{4.29}$$

for which the exact solution is:

$$u(x,t) = e^{ik(x-ut)} e^{-ik^3 \beta t} \tag{4.30}$$

The second term on the RHS of Equation (4.30) is a complex number with modulus of one. This means that the amplitude of the wave remains equal to one (no dissipation), but the speed of the wave changes from u to $u - \beta k^2$. The wave celerity is now a function of the wavenumber and the error in the wave speed is larger for large k (small wavelengths). Hence, the presence of odd-derivative terms in the truncation error makes the waves propagate at different speeds in the numerical solution, which results in numerical wiggles in the solution especially in regions where the gradients are high. This is illustrated in Figure 4.4(c).

4.5   SOLUTION METHODS FOR INCOMPRESSIBLE FLOW EQUATIONS

The solution of the governing LES equations for incompressible flows is complicated by the fact that the pressure does not have its own governing equation. The continuity equation is rather a constraint enforcing a divergence-free velocity field and is mainly used to derive a Poisson equation for determining the pressure. In most numerical methods used in LES, velocity and pressure are solved for sequentially, i.e. the momentum equations are solved first for the three (projected) velocities, and the pressure is solved subsequently from the Poisson equation; its gradients are then used to enforce a divergence-free flow field, i.e. correcting the velocities to satisfy the continuity equation. This procedure is known as the projection method and different prominent variants exist. An alternative is the method of artificial compressibility, but this is hardly used for LES. In the following the two main variants of the projection method are presented and discussed. The first variant is the fractional-step method (Chorin, 1968), which follows the above described three-step procedure, i.e. (step 1) solve momentum equations to obtain a projected (not necessarily divergence-free) flow field, (step 2) solve the Poisson equation for the pressure (gradients) using a matrix solver and (step 3) use the obtained pressure gradients to correct the velocity field from step (1). The solution of the momentum equations can be either with an explicit or implicit time-discretization method, but since the Poisson equation is of elliptic nature its solution requires a matrix solver. The second variant of the projection method is the SIMPLE (Semi Implicit Method for Pressure Linked Equation) method. Its main difference to the fractional-step method is that the projected velocity obtained in step 1 is corrected in a multi-step or iterative procedure, which is why it is called semi-implicit. Basically, the SIMPLE procedure is as follows: (step 1) advance the velocity field in time with a guessed pressure field to obtain a projected velocity field, (step 2) solve the Poisson equation for a pressure correction variable, (step 3) correct the pressure using the pressure correction variable and the velocities using the gradient of the corrected pressure, (step 4) check the updated velocity for continuity and if not fulfilled return to step 2. The SIMPLE method was originally developed for steady flows and its relatively poor convergence rates (mainly due the slow convergence of the Poisson equation) led to several, but only slightly improved schemes, e.g. SIMPLEC, SIMPLER or PISO (see Ferziger and Peric, 2002) for such flows. In LES, SIMPLE-type methods have been quite successful as the turbulent flow develops over an initial phase in which the Poisson equation does have to be solved to machine accuracy but only to a certain degree. Once the flow is fully developed, the small timesteps of LES imply that the velocity and pressure fields do not change much from one instant in time to the next. Hence the projected velocity field is already a very good estimate of the divergence-free velocity field.

The selection of the variant of the projection method depends on the chosen grid. As will be discussed below, grids with a staggered variable arrangement (called simply staggered grids) are advantageous. A staggered finite-volume grid, in which each variable is stored at a different location and has its own Control Volume (CV) is sketched in the right half of Figure 4.5. In this 2D situation, the pressure is stored in the cell center and the velocities are stored at the faces of the pressure cell. This arrangement is more complicated than the collocated grid in which all variables are stored at one location (see Fig. 4.5 left). However, solving the momentum and Poisson equations on staggered grids removes the necessity of one additional interpolation of the pressure to determine the gradients required in the momentum equations (see below). Hence, this "*strong coupling*", which also ensures the conservation of kinetic energy allows the use of the fractional-step method in contrast to the iterative procedure of the SIMPLE-type methods. However, a staggered arrangement is more difficult to implement, especially for complex geometries using body-fitted curvilinear grids (see e.g. Fig. 4.6). Grid generation and coding the Navier Stokes equations for a grid with collocated variable arrangement is much easier. However, solving the momentum equations on a

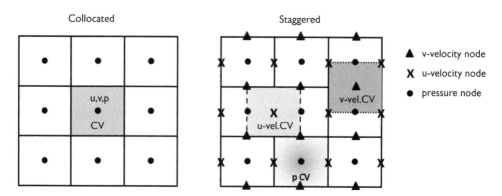

Figure 4.5 Collocated vs. staggered variable arrangement in a Cartesian grid.

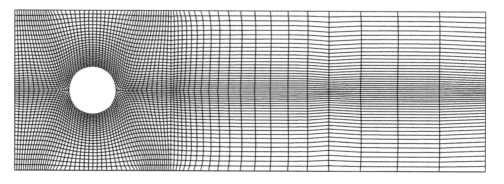

Figure 4.6 A structured, body-fitted, curvilinear grid for channel flow around a circular cylinder.

collocated arrangement requires pressure values at the CV faces, which are (linearly) interpolated from the nodal values. This additional interpolation leads to a "decoupling" of velocities and pressure and requires extra treatment, e.g. momentum interpolation (e.g. Rhie and Chow, 1983, Miller and Schmidt, 1988) to avoid unphysical oscillations in the numerical solution. The momentum interpolation adds an additional term to the Poisson equation, which destroys the energy conservation property of the solution process and requires an iterative procedure within the time step to make the flow field divergence free. The order of the space-discretization scheme is not influenced negatively by using a momentum interpolation (e.g. Melaaen, 1992) because the additional momentum interpolation term involves Δx^2, and hence a 4th order truncation error (Fröhlich, 2006). Direct comparisons of methods using staggered and collocated grids were, however, mainly carried out for steady RANS calculations and not for high-resolution unsteady LES.

Finally, matrix solvers are needed in LES of incompressible flows, either for the velocity variables when an implicit time-discretization scheme is used, but in any case for the solution of the Poisson equation for the pressure. The most efficient matrix solvers are of iterative nature and the set of algebraic equations is solved only to a certain convergence criterion. This criterion is set by the user, but mass and momentum balances need to be fulfilled to a certain degree of accuracy. What follows is that inaccuracies due to non-convergent matrix solutions can become larger than for instance due to the discretization error. However, in general the speed of the solver is more critical than its accuracy and different approaches and acceleration techniques exist. Providing details on matrix solvers is beyond the scope of the book and the interested reader is referred to standard CFD books (e.g. Hirsch, 2007, Ferziger and Peric, 2002, Versteeg and Malalasekera, 2007).

4.6   LES GRIDS

The grid provides a discrete representation of the physical domain in which the governing equations are solved. The variables are generally defined at the grid nodes in FDM, or, in case of a collocated grid, in the center of the control volumes in FVM. In the following, the four main types of grids sketched in Figures 4.6–4.10 for the flow around a circular cylinder are discussed, as well as their suitability for use in LES.

4.6.1   Structured grids

Figure 4.6 depicts a structured, curvilinear-2D grid for the flow around a circular cylinder which consists of grid lines in 2 different directions. The grid nodes are identified using indices for each direction (e.g. i for the x-direction, j for the y-direction for the 2D grid in Figure 4.6). Grid lines in one direction cross only once any grid line in the other direction so that each grid node is uniquely defined as (i, j) in a 2D grid and (i, j, k) in a 3D grid. In the FVM, the finite volume is described by the gridlines connecting the grid vertices (e.g. shaded area Figure 4.5) and has its node in the center of the CV. This results in a simple connectivity matrix and allows the utilization of efficient techniques (sparse matrix solvers) for solving the discretized governing equations. This feature is particularly important in LES and DNS because the solution of the Poisson equation is generally the most time-consuming part of the solution procedure (can take up to 80% of the total CPU time), especially for grids with a large number of points in the three directions. Depending on the shapes of the grid lines, structured grids are classified as H, O or C type (for details see e.g. Thomson et al., 1985). The discretization of the governing equations is simpler on Cartesian grids, but for complex geometries curvilinear, boundary-fitted grids are often used in flows of hydraulic interest. The main disadvantage of structured grids is the difficulty to control the distribution of the grid points in domains of complex geometry. Highly skewed and/or high-aspect-ratio grid cells are known to reduce the accuracy of the numerical solution and create convergence problems. The use of structured grids in complex domains often results in clustering of grid points in regions where this is not really needed, hereby increasing unnecessarily the CPU time. In some cases (e.g. domains with multiple bodies) it is topologically impossible to generate a structured mesh for the whole domain. Some of the problems/concerns are illustrated in Figure 4.6 for the case of a circular cylinder. The grid is refined near the cylinder to account for the large gradients there, but unfavorable grid aspect ratios and grid skewness present in the stagnation regions may lead to inaccuracies and problems with convergence. The grid is stretched away from the cylinder, and hence a greater discretization error is expected for the wake region. However, if only the near field around the cylinder is of interest, this is accepted deliberately in order to save computating time. Altogether the use of a structured grid is not ideal for this geometry. Alternatives are block-structured grids, or Cartesian grids with Immersed Boundary Method – both are introduced below. On the other hand, structured grids have been applied successfully in many LES of open channel flow in which the channel width and depth does not vary strongly.

4.6.2 Block-structured grids with matching or non-matching interfaces

For computational domains of complex shape or for domains in which the generation of a single structured grid for the whole domain results in highly skewed grid cells, block- structured grids are the best alternative to structured grids. Figure 4.7 depicts a block-structured grid for the channel flow around a circular cylinder. The original physical domain is divided into two sub-domains, for which in each a structured grid is generated. Here, an O-grid wraps around the cylinder allowing small, orthogonal cells in the vicinity of the cylinder and coarser, slightly skewed cells away from it. In the example depicted here, the O-grid is connected to a stretched H-grid at the downstream side of the cylinder, using the same number of grid points at the interface between two neighboring blocks. This is referred to as matching interface. Block-structured grids in LES require an overlap region, and the higher the order of the scheme the larger should be this region. If the number of cells of the grid on either side of the interface is different, the block-structured grid has a non-matching interface and variables have to be interpolated at the boundaries between the blocks. The use of block-structured grids with non-matching interfaces can greatly simplify the grid generation process for complex domains, allowing much more flexibility compared to the case of matching interfaces. This is of interest in LES, because it allows using local refinement strategies with very fine meshes in regions with very high velocity gradients. For the situation depicted in Figure 4.7, a very fine mesh can be used around the cylinder, resolving the boundary layer there, thus allowing for accurate predictions of drag and lift forces.

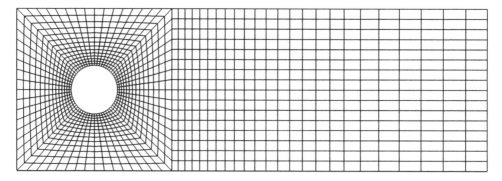

Figure 4.7 A block-structured grid for channel flow around a circular cylinder with an O-grid around the cylinder and an H-grid for the downstream part of the flow. The grid depicted here features a matching interface between O- and H-grid.

A coarser grid can then be used in the downstream section of the flow, in which eddies are convected in the direction of one set of grid lines away from the region of interest. The major drawback of non-matching interfaces of block-structured grids is the need to interpolate the variables between blocks, which violates the conservation properties. They are of particular importance in LES, as their violation tends to introduce errors of dissipative nature, and hence the quality of the LES diminishes. Finally, block-structured grids with completely overlapping blocks should be mentioned, which are known as Chimera grids. To date, such grids are used mainly in RANS simulations of flows with moving bodies.

4.6.3   Unstructured grids

An alternative to block-structured grids are unstructured grids, in which the CVs can have a variable number of neighbors (Fig. 4.8). Unstructured grids have the highest degree of flexibility in terms of the capability to generate grids for very complex domains, thereby clustering the grid nodes in regions where this is required as well as relatively rapid, but gradual, transition in cell size when moving away from the region of interest. Unstructured grids have become very popular in commercial CFD solvers due to their ability to represent any geometry. Subsequently, a number of commercial grid-generation software packages became available, with which grids can be generated rapidly and automatically, even under a certain number of constraints (e.g., minimum/maximum volume of the CV, total number of CVs in the sub-domain, maximum skewness allowed for the CVs, etc). The irregularity in the data structure is the main reason why the solvers used to invert the system of discretized equations are much slower than those for structured grids. Of particular importance for LES is the use of the Finite Volume (FV) method with unstructured grids. Even for very complex domains, the discrete conservation of mass and momentum can be exactly satisfied. The use of unstructured grids with hexahedral elements is recommended for LES, especially in regions with large velocity gradients. Based on experience with LES on unstructured grids, tetrahedral or mixed tetrahedral and hexahedral elements can introduce unphysical oscillations in the solution.

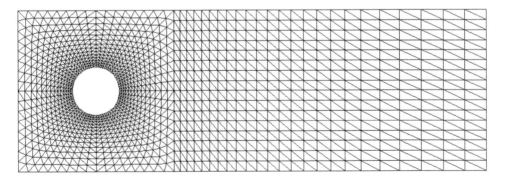

Figure 4.8 An unstructured grid for channel flow around a circular cylinder allowing for clustering a large number of elements near the cylinder.

4.6.4   Structured grids together with the Immersed Boundary Method (IBM)

The IBM method (e.g., see Mittal and Iaccarino, 2005, Iaccarino and Verzico, 2003, Fadlun et al., 2000) is another alternative of increasing popularity that allows simulations in very complex domains. Though originally proposed as a method to be used in conjunction with Cartesian or cylindrical grids, it can, in theory, also be used in any grid environment. In the IBM, the grid does not need to conform to the shape of the physical domain boundaries, and modifications are needed in the solution of the Navier Stokes equations to properly account for physical boundaries not represented explicitly by the grid. Figure 4.9 depicts the simplest and most efficient grid, i.e. a structured Cartesian grid, to be employed together with the IBM for the simulation of channel flow around a circular cylinder. The physical domain boundaries of the cylin-

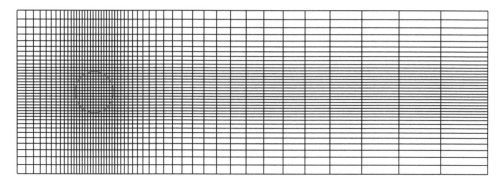

Figure 4.9 A structured, Cartesian grid for channel flow around a circular cylinder using the immersed boundary method.

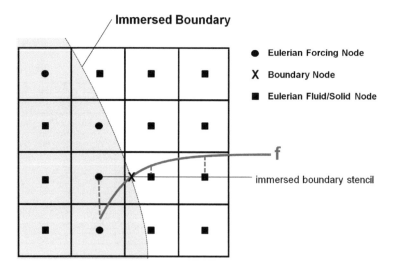

Figure 4.10 Illustration of the immersed boundary method.

der are immersed or embedded within the grid and the no-slip boundary condition of the cylinder requires extra treatment. The general idea is to account for the presence of the boundaries of an object in a flow domain by adding forcing terms in the governing flow equations in the vicinity of the immersed boundary. An example of a typical immersed boundary treatment is illustrated in Figure 4.10. The body is immersed in the Cartesian grid and the velocity of the fluid on the boundary is known, i.e. through the no-slip condition. The no-slip condition is imposed on the surrounding fluid by adding a force term to the momentum equations, in Figure 4.10 at the first grid node inside the immersed boundary. The magnitude of the force is such that it enforces a pre-determined target velocity for that node, and the target velocity is determined through a reconstructed velocity field using the values from the neighboring (fluid) grid nodes and the no-slip velocity on the immersed boundary. Several interpolation schemes are available and detailed reviews of the different approaches are provided by Iaccarino and Verzicco (2003) and Mittal and Iaccarino (2005).

The IBM method is particularly powerful for simulations with moving boundaries due to its simplicity and accuracy, which makes it very attractive for LES of such flows. In such simulations, the underlying (Cartesian/cylindrical) grid is stationary, which ensures the observation of conservation principles, and at the same time allows the use of fast and efficient sparse matrix solvers. The IMB has been successfully used in a number of simulations of hydraulic interest, for instance for the flow around artificial submerged vegetation (Stoesser et al., 2009), the flow in a meandering channel with large bedforms (Kang et al., 2011) or the flow through a porous bed. These will be introduced in the Application Chapter 9.

Chapter 5

Implicit LES (ILES)

5.1 INTRODUCTION

As already mentioned in section 4.1, the numerical dissipation provided by the scheme used to discretize the differential terms in the governing equations can remove energy from the resolved scales, similar to a classical subgrid-scale model. This forms the basis of another type of LES methods that do not use an explicit subgrid-scale model but rather rely mostly on the numerical dissipation resulting from the discretization of the convective terms in the momentum equations. The governing equations before discretization are formally the unfiltered Navier-Stokes equations, like in DNS. The truncation error of the discretization (as discussed in section 4.2) is then used to model the effects of the unresolved scales, hence acting as an implicit SGS model. This type of methods is therefore generally referred to as implicit LES or ILES (e.g. see Boris et al., 1992, Adams et al., 2004). It should be noted, however, that not all dissipative discretization schemes are appropriate for ILES. The reason is that, as discussed in section 4.3, most schemes (e.g. classical upwind-biased schemes) remove substantial energy from the whole resolved wave-number range rather than removing energy only from the smallest resolved scales, and the amount of energy locally removed from the flow is not determined by the eddy content of the instantaneous flow fields but rather depends on the local grid spacing and numerical discretization.

The goal of the present chapter is to provide only a brief description of the ideas behind the main types of ILES methods without providing the details on these. For a comprehensive overview of the standard methods and sample applications the reader is referred to the recent book on ILES by Grinstein et al. (2007). This book also includes a new type of ILES method called Adaptive Local Deconvolution Model (ALDM) developed by Hickel et al. (2006), which is introduced briefly in section 5.3.

It should be recalled that in classical LES the governing equations are the spatially filtered/averaged Navier-Stokes equations (2.10, 2.11) containing a subgrid-scale stress term τ_{ij}^{SGS} which is modeled by an explicit SGS model. Numerical aspects did not enter in the derivation of these equations, but as the equations have to be solved numerically, numerical effects and in particular truncation errors influence the solution unless the grid size is much smaller than the characteristic flow scales. Hence, as already mentioned in section 4.1, there is generally an interaction between the explicit SGS model used and the numerical aspects such as discretization and grid resolution and the effects of the two cannot be delineated. In contrast, in ILES there is no explicit model introduced for τ_{ij}^{SGS} and the (mainly damping) effect of the unresolved scales on the resolved motion is entirely accounted for by the truncation errors introduced by the discretization scheme.

The success of the ILES method depends on the discretization scheme used for the nonlinear convective fluxes (simple schemes were introduced in section 4.2.2) which controls the amount of numerical dissipation added to the solution. One common approach in the past was to use standard (pre-existing) higher-order-upwind discretizations or even more complex non-oscillatory shock-capturing schemes involving flux limiters. The latter type of schemes can capture high gradients in the resolved velocity and scalar fields without introducing nonphysical oscillations in the numerical solution in regions where sharp variations in the resolved variables develop. The Monotonically Integrated LES (MILES) scheme discussed briefly in section 5.4 is of this type and in addition to smoothing the solution the numerical dissipation introduced through the truncation error acts as a physically-motivated SGS model. Such schemes act as a filtering mechanism for the small scales, but were shown to result in many cases in excessive damping of the smallest resolved scales and inaccurate predictions of the mean flow and turbulence statistics due to their poor capacity in reproducing the effects of the unresolved scales (e.g., see Garnier et al., 1999). In other words, for many of these schemes the truncation error acts as a poor SGS model. From a purely numerical perspective, these discretizations add numerical dissipation with the aim to stabilize the numerical simulations. This is the main reason why ILES was for a long time not very popular with the LES community which viewed ILES as simply LES simulations with added artificial (numerical) dissipation whose magnitude is controlled by the grid and discretization rather than by the flow physics. However, recent efforts have focused on proposing discretization schemes that are specifically designed and optimized such that the truncation error behaves effectively like a physically-motivated SGS model. The Adaptive Local Deconvolution Model (ALDM) introduced briefly in 5.3 is of this type. It is based on the partial and approximate reconstruction of the complete (resolved and unresolved) velocity field by the method of deconvolution (or inverse filtering) as introduced already in section 3.4.3. This method provides a framework for systematic derivation of a discretization scheme with the desired properties.

Most ILES methods, including ALDM, use nonlinear stable discretization schemes. By nonlinear we mean schemes in which the coefficients are functions of the velocity field at the previous time level rather than constants. This is different from the linear (explicit) discretization schemes discussed in section 4.2, where the velocity in a cell at the new time level can be expressed as a linear combination of the velocities at the neighboring cells at the previous time level and the coefficients in the linear combination are constants.

5.2   RATIONALE FOR ILES AND CONNECTION WITH LES USING EXPLICIT SGS MODELS

One important observation that holds both for LES with explicit and implicit modeling is that the exact equations satisfied by the numerical solution do not coincide with the original exact equations satisfied by the continuous solution (e.g., see discussion in Hickel et al., 2006 and also the simple example discussed in section 4.4 for the linear pure advection equation). The former equations are called the modified equations.

In ILES, the modified equations do not contain an explicit SGS term resulting from spatial filtering or averaging. However, the usually applied finite-volume discretization of the governing equations for the unfiltered variables introduces trunctation errors (see 4.2.1 and 4.2.2). Hence, in ILES the modified equation for the control-volume averaged velocity vector, $\bar{u}_N$, contains a truncation-error term, but no SGS residual term. In contrast, in classical LES with explicit SGS modelling as discussed in Chapter 3 the modified equation satisfied by the spatially-filtered velocity contains a SGS residual term and generally also a truncation error term, which should however (ideally) not dominate.

The main idea in ILES with a physically-motivated implicit SGS model is to cast the truncation error introduced by the discretization of the non-linear (convective) terms in the Navier-Stokes equations as the divergence of a tensor:

$$T_i^C \cong \frac{\partial}{\partial x_j}(\tau_{ij}^I) \qquad (5.1)$$

where, for a control volume, T_i^C is the component of the error vector $\overline{T}^C$ in the i direction and τ_{ij}^I plays the role of the SGS stress tensor τ_{ij}^{SGS} in LES with explicit SGS modelling. Thus, ILES can be formally equivalent to classical LES using explicit SGS modelling if the truncation error term has properties that are similar to those of an explicit SGS term. This automatically implies that not all discretizations of the convective terms in the governing equations result in expressions for $\overline{T}^C$ that can be written in the form given by equation (5.1). Thus, a first requirement is that the truncation-error vector can be written as the divergence of a second-order tensor, τ_{ij}^I as in Equation (5.1). For some popular ILES methods (e.g., FCT, MPDATA, ALDM – see 5.3 and 5.4 below), one can further show (e.g., Grinstein et al., 2005, Margolin et al., 2006) that the expression for τ_{ij}^I contains a tensor-valued eddy-viscosity (dissipative) term, similar to a tensor version of the classical Smagorinsky SGS model, and a term similar to a scale-similarity SGS model. If the aforementioned condition is satisfied, the truncation error effectively behaves as an implicit SGS model, more precisely like that given by a mixed SGS model (as described in section 3.4.2) in the example given.

Moreover, in ILES one tries to obtain via specifically designed nonlinear discretization schemes of the convective terms a truncation error in the modified equation that models the proper energy transfer between resolved and unresolved scales on the grid on which the equations were discretized. The truncation error from the viscous term can be neglected if high-order centered schemes are used to discretize the diffusive flux (see also discussion in Hickel et al., 2006).

5.3 ADAPTIVE LOCAL DECONVOLUTION MODEL (ALDM)

The main idea of ALDM developed by Hickel et al. (2006) is to design a novel non-linear discretization scheme with free parameters that allow controlling the truncation error and are calibrated so that this acts as a physically motivated SGS model.

The method employs the finite-volume concept of Schumann (1975) introduced in section 2.4, that is the resolved quantities are the control-volume averaged values and explicit filtering is not introduced, but the approach corresponds to the use of a top-hat filter. The volume-balance method of Schumann's approach introduces fluxes at the control-volume faces which are averages over the local (or unfiltered) values involving unresolved quantities due to subgrid-scale fluctuations. These local values corresponding to the unfiltered solution are approximated from resolved quantities by the method of approximate deconvolution with the aid of solution-adaptive deconvolution polynomials. Face averaging and relating face values to cell-average values is then performed with an ALDM specific flux function, an operation called flux reconstruction which yields the desired convective flux into or out of the control volume. The non-linear deconvolution operator and the flux function contain free parameters which can be used to adjust the spatial truncation error of the entire discretization scheme. It should further be noted that this error can be easily written as a divergence of a tensor as in Equation (5.1). In a second step of the ALDM development, the discretization parameters inherent to the solution-adaptive discretization scheme and the flux function used for the convective term are determined in such a way that the spatial truncation error of the method acts as a physically-motivated SGS model which adds numerical viscosity only in the regions where the small scales are energetic. This is the main difference to standard approaches to construct non-linear discretization schemes in which the parameters are determined such that the highest order of accuracy is obtained.

The extension of ALDM for transport of passive scalars is discussed by Hickel et al. (2007). The free parameters in the discretization of the momentum and scalar transport equations are calibrated in such a way that the model matches as closely as possible the requirements of turbulence theory in freely decaying isotropic turbulence at high Reynolds and Peclet numbers, more precisely the analytical expression for the eddy viscosity and eddy diffusivity. Once determined, the model parameters are kept unchanged. Thus, ALDM does not require additional calibration of the model parameters for complex flows, which is a major advantage. For incompressible flows, the computational overhead of using ALDM is comparable to that of the dynamic Smagorinsky model with a second-order centered scheme.

ALDM simulations of large-scale forced and decaying 3D isotropic turbulence (Hickel et al., 2006) showed very good agreement with theory and experimental data. The model was also successfully applied to predict the instability growth and transition to developed turbulence in transitional flows (e.g., see Hickel et al., 2006 and 2010 for the 3D Taylor Green vortex problem, and Hickel and Adams, 2008 for transition in a Blasius zero-pressure-gradient boundary layer). The model extension for wall bounded flows (e.g., the use of a van Driest damping function is recommended) is discussed in Hickel and Adams (2007). ALDM simulations of turbulent channel flow at Reynolds numbers defined with the wall friction velocity and the channel half depth (Re_τ) of up to 950 yielded slightly better results than the dynamic Smagorinsky model at the same grid resolution. Hickel and Adams (2008), Hickel et al. (2008) and Meyer et al. (2010) discuss the performance of the model for three more complex cases of attached and separated flows (turbulent boundary layers with adverse-pressure-gradients, turbulent channel flow with streamwise periodic constrictions and flow past a cylinder).

Again the ALDM results were shown to compare well with experiment and results of well-resolved explicit LES. ALDM was also validated for turbulent transport of passive scalars in forced isotropic turbulence and turbulent channel flows (Hickel et al., 2007) and extended and applied to stratified flows (Remmler and Hickel 2012). The fact that ALDM predicts wall bounded separated flows and scalar transport with an accuracy that is similar to that of a dynamic Smagorinsky model on meshes with similar resolution, makes it a good candidate for hydraulic applications.

5.4 MONOTONICALLY INTEGRATED LES (MILES)

A detailed review of Monotonically Integrated LES (MILES) methods and a discussion of their capabilities to predict various types of simple and complex turbulent flows can be found in Grinstein and Fureby (2004) and in the recent book of Grinstein et al. (2007).

MILES is based on the use of a mixed discretization scheme in which the convective flux at control volume faces is calculated with a high-order (at least second order), non-dissipative scheme when the solution is smooth and with a low-order (upwind), dissipative scheme when the solution tends to develop sharp gradients. With this mixed scheme, the convective flux can be written as

$$(uf) = (uf)^H - (1 - \Gamma)((uf)^H - (uf)^L), \tag{5.2}$$

where $(uf)^H$ is the high-resolution flux function of high order and $(uf)^L$ is a low-resolution dispersion-free but dissipative flux function that is well-behaved near sharp gradients (Grinstein et al. 2005). Γ is a non-linear flux limiter function which controls the contribution of the dissipative flux and hence the gradients of the resolved variables such as to prevent the appearance of numerically generated oscillations (Hirsch (2007). The proponents of MILES refer to the entire mixed scheme as non-linear high-resolution scheme, but this should not be confused with high-resolution LES that refers to LES performed on very fine meshes. The goal of the limiter is to restrict the flux of the conserved quantity into a control volume to a level that will not produce a local maximum or minimum of that quantity in the control volume (Ferzinger and Peric, 2002). The use of the high-resolution flux fluctuation alone would, in the absence of the damping effect of an explicit SGS model, lead to oscillations and possible divergence if sharp variations occur in the solution, as would be particularly the case when turbulence is strong and grid resolution low. The application of the flux limiter stabilizes the solution by providing attenuation of fluctuations/wiggles occurring at length scales close to the local grid spacing (smallest resolved scales). Thereby, numerical dissipation is introduced, particularly in regions where large fluctuations develop, and this is the implicit SGS-model effect of the MILES method. In regions where the solution is smooth, the flux limiter function Γ is equal to 1 and the high-resolution flux function is used in (5.2).

As already pointed out, not all flux-limited schemes perform well when used for LES. To provide a strong attenuation of the small scale fluctuations developing into the resolved velocity and scalar fields, the truncation error should be expressible in divergence form (see Equation 5.1) and the leading order term of the truncation error should be dissipative (see also section 4.4).

One popular choice for the flux functions in Equation (5.2) (Fureby and Grindstein, 2002) is to base $(uf)^H$ on linear interpolation (this results in the second-order cell-centered CDS scheme introduced in section 4.2.1 as used in standard non-dissipative LES codes) and $(uf)^L$ on upwind-biased schemes (see section 4.2.1 and Hirsch, 2007). An analysis of the truncation error of the convective terms in the modified equation (Fureby and Grindstein, 1999, 2002, Rider and Margolin, 2003) showed that the two leading order terms have a dissipative and, respectively, a dispersive character (see section 4.4). Similar to ALDM, the dissipative term can be expressed as a tensor-valued eddy-viscosity SGS model (see Equation 5.1). The dispersive term has a form similar to that of a higher order scale-similarity SGS model that can be obtained using approximate deconvolution of the resolved velocity field (see section 3.4.3). Thus, one can argue that the leading order terms in such models behave similar to a mixed explicit SGS model as described in section 3.4.2, so that the truncation error of the convective-term discretization should act as a physics-based implicit SGS model, if the flux limiter Γ is chosen accordingly.

The flux limiter function, Γ, used to express the fluxes at the faces of the control volume in a mixed scheme is a scalar function that depends on the local velocity field so that the limiter is non-linear. The analytical constraints that need to be satisfied by Γ used in mixed, so-called high-resolution schemes and popular expressions for Γ are discussed in detail by Hirsch (2007). In terms of the choices for Γ, the main difference is between methods in which the 1-D flux limiter can be formulated in terms of the ratio of successive velocity gradients for neighbouring meshes $r = (u_i^n - u_{i-1}^n)/(u_{i+1}^n - u_i^n)$ and methods in which r is a function of the velocity at nodes $i - k$ to $i + k$ ($k > 1$), both at the time level at which the solution is known. In the former category, some of the most popular choices for MILES are the minmod limiter, the superbee limiter, the van Leer limiter, and the GAMMA limiter (Hirsch, 2007). The very popular TVD scheme is obtained for a particular value of the free model parameter in the definition of the GAMMA limiter. The also popular Flux-Corrected Transport (FCT) (Fureby and Grinstein, 2002) and the Piecewise Parabolic Method (PPM) (see Colella and Woodward, 1984) schemes fall into the second category. The generally better performance of these schemes for LES simulations is explained by the fact that they are less diffusive compared to the schemes for which $\Gamma = \Gamma(r)$. Based on simulations of turbulent channel flow Grinstein et al. (2005) concluded that among the schemes for which $\Gamma = \Gamma(r)$, the GAMMA limiter gives the most accurate predictions. The accuracy of the solution was comparable to the one given by the FCT method. Grinstein et al. (2005) also discuss the extension of the FCT method for 3D simulations using multi-dimensional flux limiting rather than employing the 1D FCT limiter in each direction. Such an approach was shown to produce more accurate results especially for incompressible flows.

The performance of MILES for simple and complex flows, including massively separated flows in complex geometries, is discussed in details in the book by Grinstein et al. (2007). The review paper by Grinstein and Fureby (2004) gives relevant examples of application of MILES for wall bounded flows, free shear flows with transition, separated flows past bluff bodies and dispersion of pollutants in urban environments. The last two types of applications are directly relevant for hydraulic engineering problems.

One criticism of MILES-based methods compared to ALDM is that the spatial truncation error of the discretization is not optimized such that it models correctly the energy transfer between resolved and unresolved scales. Further efforts are needed to propose a general framework to control the properties of the implicit SGS model that results from a particular choice of the discretization scheme. This should eventually allow optimizing the parameters (e.g., shape of the flux limiter) in these schemes to further improve the capability of MILES to predict complex turbulent flows.

Chapter 6

Boundary and initial conditions

Large Eddy Simulations are carried out in finite-size computational domains chosen by the user. For solving the governing differential equations, boundary conditions must be specified at all boundaries of this domain, as well as initial conditions for the dependant variables within the entire domain at the start of the time-marching simulation. The computational domain depends strongly on the geometry and flow conditions of the problem considered, and a variety of examples will be given in the Applications Chapter 9. The boundaries of the computational domain can be of various types, and those occurring in hydraulics problems are illustrated by way of the calculation domain being a stretch of a river in Figure 6.1. The bed consisting of a solid wall (which in exceptional cases may be permeable) and the free surface are physical boundaries. The wall boundary at the bed may be horizontal or inclined, but in channels with rectangular cross section and in the presence of man-made structures or idealized vegetation elements, vertical walls are also present as physical boundaries. In addition, artificial boundaries are usually introduced in order to limit the size of the calculation domain. Considering only a stretch of a river as shown in Figure 6.1, an inflow and an outflow boundary needs to be chosen by the user limiting the domain in the streamwise direction, but when the flow can be considered statistically homogenous in one of the other directions, the domain can be limited in this direction by introducing artificial periodic boundaries. Often the conditions at the boundaries cannot be formulated to represent exactly the real physical conditions prevailing at the boundaries, e.g. when details of a rough wall are unknown or velocity fluctuations at the inflow or at a free surface need to be approximated. Hence models are necessary for these approximations and these are named super-grid models. The uncertainties and possible errors introduced due to such super-grid modelling can be orders of magnitude larger than those due to subgrid-scale modelling and hence the quality of a LES is generally greatly affected by the treatment of each boundary in the numerical simulation.

The specification of boundary conditions depends on the numerical procedure employed. In finite-volume methods conditions must be provided that allow the evaluation of the convective and diffusive fluxes at the faces of the numerical control volumes coinciding with boundaries in the discretized filtered Navier-Stokes equations. This requires the specification of either the fluxes or the values of the dependent variables at the boundaries, or a means to express these as a function of interior data. This amounts to either specifying values at boundaries, e.g. all three velocity components at the inflow plane, called Dirichlet condition, or the specification of

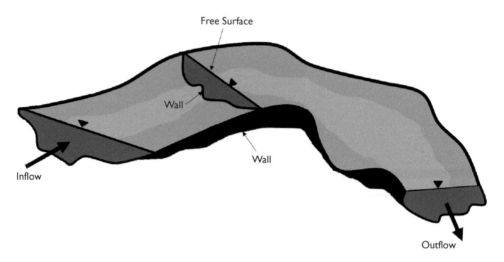

Figure 6.1 Computation domain for an open-channel flow and types of boundaries occurring.

fluxes generally involving gradients, such as at the outflow or at walls, in a Neumann condition. A third possibility is to apply periodic conditions when periodicity of the statistical quantities in certain directions can be assumed. This approach is popular and often employed as a physically more realistic alternative to inlet/outlet conditions, especially when details of the flow at the inlet/outlet are needed to ensure accuracy of the calculations in a domain but are not known, and in cases with homogeneity of the flow in certain directions to keep the size of the computation domain in these directions relatively small.

6.1   PERIODIC BOUNDARY CONDITIONS

Periodic conditions can be used at artificial boundaries when the flow is statistically homogenous in a certain direction or the geometry is periodic in one or two directions. The first case exists in developed open channel flow (Fig. 6.2). Here periodicity in the streamwise direction prevails as the distribution of statistical quantities over the cross section is the same at each cross section. In wide open channels, i.e. without the influence of side walls and in the absence of secondary motions, the flow is homogenous in the spanwise direction and hence periodicity can be assumed also in this direction. Further, for instance in cases with dense emergent vegetation, bottom friction is negligible and the numerical simulation resolving the flow around the individual vegetation elements is to good approximation homogenous in the vertical direction. Hence not the entire depth needs to be covered in the calculation but only a slice assuming vertical periodicity which allows to reduce considerably the number of grid points required.

The flow over a cube matrix as shown in Figure 6.3 exhibits geometric periodicity in stream- and spanwise direction. When the vertical planes bounding the computational

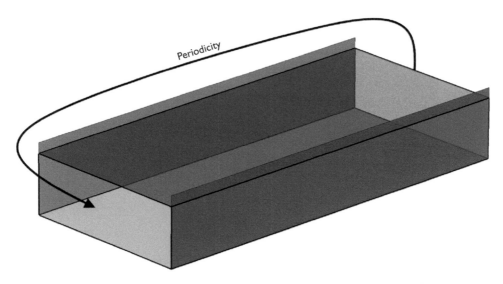

Figure 6.2 Flow through a straight open-channel with statistical homogeneity in the streamwise direction.

domain are chosen such that they lie in the middle between consecutive cubes in either direction (Fig. 6.3), then the statistical quantities are the same on opposite planes and hence periodic conditions can be applied. The flow between the planes is in this case of course not homogenous and the instantaneous flow is not periodic; it is characterized by turbulent structures, the largest of which carry the biggest amount of kinetic energy and hence these large structures are the greatest contributors to the flow statistics. This implies that the extent of the computational domain at whose boundaries periodicity is enforced has to be selected with great care. If the distance between streamwise periodic boundaries is considerably smaller than the size of the largest turbulence structures occurring in the flow, such structures will be artificially confined within the domain, which leads to an unphysical flow behaviour and hence to erroneous statistics. Therefore the proper choice of distance between periodic boundaries is essential in an LES that uses periodic boundary conditions. For an open-channel flow over a smooth bed the minimum distance between streamwise periodic boundaries, l_x, can be estimated to be six times the water depth, h, assuming that the longitudinal extent of the largest turbulence structures is typically less than three times the water depth. The spanwise distance between periodic boundaries in the cross-streamwise direction is not as critical as that in the streamwise direction; however structure lock-in in the spanwise direction should be avoided by proper choice of the spanwise distance, l_y, which can be estimated to be at least twice the water depth. Using periodicity in the vertical direction in emergent vegetation flow, the vertical extent of the domain should be of the order of the distance between vegetation elements in order to resolve all relevant turbulent structures.

The numerical treatment of periodic boundaries is such that on both ends of the simulation domain so-called ghost cells are added to the domain and the variables at one side of the domain are copied after every computed time step into the ghost cells

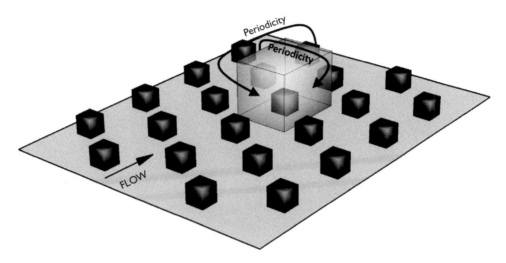

Figure 6.3 Flow over a matrix of cubes that exhibits geometric periodicity in stream- and spanwise direction – the shaded area is an example of a preselected computational domain.

of the other side and vice versa. In spanwise and vertical directions, all variables i.e. the three velocity components and the pressure are exchanged, which requires no further treatment. In the streamwise direction a pressure gradient is required between upstream and downstream end that balances the shear stresses acting on the walls. Usually, pressure values are also exchanged and an external force that drives the flow is added to the filtered momentum equations as a source term. The magnitude of the force is chosen to ensure a constant mass flux.

6.2 OUTFLOW BOUNDARY CONDITIONS

As mentioned in the introduction to the chapter, the outflow boundary in an LES is an artificial boundary and has to be placed as far downstream of the region of interest as possible. However, the relatively high resolution requirements of LES constrain the distance between the region of interest and the outflow boundary so that the most important condition to be satisfied is that the region of interest is not affected by the artificial conditions imposed at this boundary. The elliptic nature of the Navier-Stokes equations on the other hand implicates that, strictly, the outlet boundary conditions affect the flow in the upstream direction, but as most flows in hydraulic engineering are convection dominated, in effect the values at the boundary have only minor influence on the solution inside the domain. For such flows, the easiest way to approximate the values at the outlet boundary is to assume zero gradients along streamwise gridlines, which is an extrapolation of 0th order. For the convective terms this treatment yields:

$$\partial u_i / \partial x_i = 0 \qquad (6.1)$$

Diffusive fluxes are approximated with one-sided differences. When in situations with obstructions causing larger and persistent eddies, the outflow boundary cannot be placed sufficiently far downstream, application of the condition (6.1) is not straight forward. In this case eddies are convected to the outlet and may result in negative velocities there, so that instantaneously fluid enters the domain. This scenario would lead to a negative pressure gradient near the outlet and hence to numerical oscillations which can travel upstream and destabilize the flow in the interior of the domain. Damping of such pressure oscillations can be achieved by grid stretching or by artificially increasing the viscosity in the vicinity of the outlet. With that, eddies are artificially stretched and/or weakened and possible pressure oscillations are decreased.

A more physically reasonable and now mostly used alternative to damping the turbulence near the outlet is to employ a convective boundary condition. This condition requires solving an unsteady 1D convection equation along streamwise gridlines i.e.

$$\partial u_i / \partial t + U_{conv} \partial u_i / \partial x_i = 0 \qquad (6.2)$$

This equation is solved for the velocity components on the boundary using a first order backward difference scheme to compute the spatial derivative and an explicit discretization scheme for the time derivative. The convective velocity U_{conv} can be set in such a way that global mass conservation is achieved. The convective boundary condition is today the standard and preferred outlet condition for LES and has been found to work well in many applications including flow around a circular cylinder (e.g. Breuer, 1998a), flow over a backward facing step (Le et al., 1997), or the flow over a hill (Garcia-Villalba et al., 2009). The convective condition is used in all application examples presented in the Applications chapter of this book, except when periodic boundary conditions are applicable.

6.3 INFLOW BOUNDARY CONDITIONS

In Large Eddy Simulations the inflow boundary is usually an artificial boundary at which values of the quantities to be computed, i.e. u_i and p, have to be specified. As mentioned above, flows in hydraulics are convection dominated – so that the values specified at the inflow boundary influence greatly the values inside the calculation domain. As a consequence, in open-channel flow simulations physically realistic velocity and pressure values are needed at the upstream end of the domain. The easiest and by far the most common treatment of the inlet boundary is to prescribe the velocity distribution over the inlet area as a Dirichlet boundary condition and to extrapolate the pressure from the inside to the inlet plane. However, in contrast to RANS calculations not only the time-averaged velocity distribution is required but also the specification of physically realistic velocity fluctuations is needed. This poses a significant challenge in Large-Eddy Simulations, since high resolution (spatially and temporally) velocity data at the inlet plane of the flow domain, for instance from experiments, is rarely available.

In some situations, where the approach flow is uniform and virtually free of turbulence, or when the turbulence and large-scale flow structures are a result of flow separation inside the calculation domain, prescription of a time-averaged velocity profile at the inflow boundary may be sufficient. Examples for this are the flow over a backward facing step (Fureby, 1999; Le et al., 1997) or the flow around long square and circular cylinders (Rodi et al., 1997, Breuer, 2000). In the first 2 cases the location of flow separation is predefined through the geometry (sharp corners) and hence does not depend on the upstream flow conditions. However, in cases in which the approach flow carries turbulence and its conditions influence the downstream flow, neglecting the turbulence at the inflow could lead to significant errors or to extremely long and hence resources-consuming development lengths. For instance, the location of flow separation on curved surfaces is known to be influenced by the upstream turbulence. Garcia-Villalba et al. (2009) calculated the flow over a three-dimensional mildly sloped hill and stressed the importance of matching the turbulent inflow conditions to the experimental ones in order to accurately predict the location of flow separation on the hill.

Recently, considerable research has been dedicated to generating adequate inflow boundary conditions for LES. This issue is also important in Hybrid LES/RANS methods covered in Chapter 7 at the transition from RANS to LES zones/regions but will only be dealt with here.

The most common and practical methods for generating inflow conditions for LES will be discussed in the following.

6.3.1   Precursor simulations

For channel-type flows, realistic inlet conditions can be obtained from pre-cursor simulations of an upstream placed sub-channel, the geometry and wall boundary conditions of which match those at the inlet of the actual calculation domain (Fig. 6.4). The sub-channel flow is assumed to be developed so that the boundary conditions are

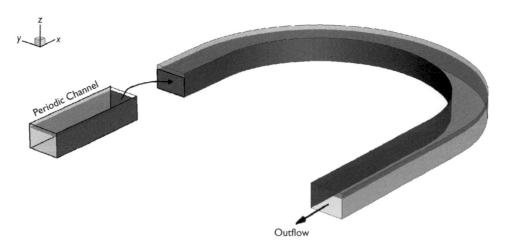

Figure 6.4 Pre-cursor simulation performed for an upstream periodic channel to generate realistic inflow velocities for the flow in an open-channel flow through a 180 degree bend.

chosen to be periodic in the streamwise direction. Once the flow field has developed into a fully turbulent and statistically steady state, the time-varying velocity data at one cross-section are copied to the inlet plane of the main simulation. Ideally, not only the geometry, but also the grid and temporal resolution match exactly the conditions at the inlet plane. If not, interpolation of velocity data in space and/or in time is required, which, especially if performed in time, could lead to unphysical turbulence spectra. One possibility is to perform the pre-cursor simulations before the actual simulation and write the results for one plane into a file from which the main simulation then reads the data as inlet boundary condition. This can be a challenge in terms of storage due to the high resolution requirements in space and time of the pre-cursor simulation which should cover the same physical time as the actual simulation. If less physical time is stored, the precursor results need to be fed-in repeatedly. A further possibility is to run the pre-cursor simulation calculations parallel to the main simulation and to feed-in the results of the periodic sub-domain directly as inlet boundary condition of the main simulation. This treatment inevitably leads to an increase in the required computational resources of the entire simulation. In cases in which the straight-channel pre-cursor simulation does not match the desired or (statistically) known inlet conditions of the main simulation, the use of forcing terms and rescaling of turbulent fluctuations are adequate measures to better approximate the inlet conditions. Forcing and rescaling has been done successfully by Garcia-Villalba et al. (2009) who rescaled velocities and turbulent fluctuations of a closed smooth channel to match closely the profiles of mean velocity and turbulent kinetic energy of thick boundary layer that was artificially created at the upstream end of a wind tunnel experiment.

6.3.2   Time-averaged velocity profile superimposed with synthetic turbulence

Time-averaged velocity profile superimposed with random numbers (or white noise)

The easiest but least successful method of generating a turbulent inlet velocity field is super-imposing white noise on the mean velocity field. Though the random numbers can be scaled to match experimental *rms* values, the resulting velocity signal is not physically realistic as the spectrum of the prescribed fluctuations does not exhibit a decay towards higher frequencies as is the case in real turbulence. More importantly, almost immediately downstream of such an artificially created velocity signal, the high frequency fluctuations are damped quickly by the numerical method demanding a divergence free flow field. This influences negatively the low frequency more energetic fluctuations. As a result, the flow requires a certain approach flow length over which the velocity signals adjust themselves towards physically realistic turbulence. This length is hard to estimate *a-priori* and depends on the geometry and Reynolds number. In LES, every grid point that can be saved by avoiding a long "adjustment length" is important so that in simulations in which the proper upstream conditions are important, alternative, more advanced methods should be favoured.

Several studies have addressed the poor performance of the use of random noise superimposed on a velocity profile by generating more realistic turbulence for the inlet plane. The idea of generating synthetic turbulent fluctuations to be superimposed onto a time-averaged velocity profile makes use of Taylor's frozen turbulence hypothesis, in order to relate spatial and temporal turbulence. By using an appropriate convective velocity it can be assumed that turbulence statistics such as turbulence intensities and Reynolds stresses from spatial simulations are similar to those from temporal simulations.

Method of Lee et al.

Based on this assumption, Lee et al. (1992) suggested generating stochastic turbulent fluctuations with a prescribed energy spectrum from inverse Fourier transformations. This can be accomplished by providing two transversal wave numbers, the frequency of the spectrum and random phase angles in the streamwise direction. The use of Lee et al.'s (1992) procedure ensures that the resulting signals do not contain excessive small-scale motions which would have resulted if simply random numbers were used to generate the velocity fluctuations. For the flow over a backward facing step, Le et al (1996) superimpose a mean velocity profile with fluctuations with a prescribed energy spectrum following the method suggested by Lee et al. (1992). Because of the inhomogeneity in the wall-normal direction, calculated fluctuations are rescaled to conform to the three normal Reynolds stress components and the wall-normal shear stress associated with a turbulent boundary layer profile. As mentioned above, the incompressibility condition must also be considered, which implies that the generated inflow fluctuations must be zero when summed over the inflow plane. Though fairly realistic velocity signals were achieved, a certain approach length is still required for the flow to develop into physically correct turbulence; this approach length however is considerably shorter than the one needed if white noise was superimposed at the inlet.

In an effort to improve the velocity signals at the inlet plane the fluctuations at each point can be manipulated, for instance through the phase angle to compute the Fourier coefficient, to attain specific spatial correlations expressing the coherence of a turbulent flow (Kondo et al., 1996).

Digital filter method

Klein et al. (2003) proposed a general and efficient method for generating artificial turbulent inflow conditions for spatially inhomogeneous flows based on a digital filtering approach. The digital filtering approach does not assume flow similarity or equilibrium.

In the first step of this approach, a provisional velocity field u_m is generated for each velocity component from a random data series $r_m(m = 1 \dots k)$, which possesses prescribed two-point statistic and zero mean ($\overline{u}_m = 0$), using a digital filter defined by:

$$u_m = \sum_{n=-N}^{N} b_n r_{m+n} \tag{6.3}$$

where b_n's are the filter coefficients that have to be determined and N is the extent of the filter support (filter size). The random data series r_m has to satisfy zero-mean ($\overline{r_m} = 0$) and unit variance ($\overline{r_m r_m} = 1$). With this, one can show that the following relation holds between the two-point correlation function of u_m and the filter coefficients b_j (in 1D – for x-direction):

$$R_{uu}(k\Delta x) = \frac{\overline{u_m u_{m+k}}}{\overline{u_m u_m}} = \frac{\displaystyle\sum_{j=-N+k}^{N} b_j b_{j-k}}{\displaystyle\sum_{j=-N}^{N} b_j^2} \qquad (6.4)$$

where $k\Delta x$ is the distance from the reference point considered and Δx is the grid spacing in x. For a given two-point correlation R_{uu} the filter coefficients b_j can be determined by inverting relation (6.4). Klein et al. (2003) prescribe the two-point correlation by assuming a Gaussian shape and by specifying the integral length scale L_x. This is consistent with homogeneous turbulence, but other correlation shapes can be used. The filter size N is chosen to be twice the ratio of length scale to grid spacing ($N = 2L_x / \Delta x$). A 3D filter is then obtained by the convolution of three 1D filters.

In the second step, the time series from the first step having zero mean, unity variance and zero cross-correlations are modified using a linear transformation (Lund et al., 1998), such that one obtains new velocity time series with desired first (mean values) and second order point statistics (correlations between the different velocity components). The single point statistics are generally obtainable from experiment or from RANS in a hybrid method.

The afore-mentioned procedure allows obtaining instantaneous velocity data in the inflow plane with appropriate spatial correlations from an initial set of 3D random data. The method was validated by Klein et al. (2003) for two cases (DNS of plane turbulent jet, 2D DNS of primary break up of a liquid jet) in which the turbulence scales were spatially uniform over the entire inlet plane.

Veloudis et al. (2007) proposed a new version of the digital filtering approach developed by Klein et al. (2003) that allows for spatial variation of the input turbulence length scales, instead of assuming a single prescribed spectrum at all inflow points, at a reasonable computational cost. The increased accuracy of the modified method using spatially varying turbulence length scales was demonstrated for the case of channel flow with a periodically repeating constriction. In particular, the prediction of the turbulence profiles was found to be more accurate compared to the case when a constant turbulence length scale was assumed. The use of the Fast Fourier Transform for the 3D filtering process resulted in significantly higher computational speed.

A computationally very efficient digital-filter-based approach for generation of turbulent inflow conditions for spatially developing flows was proposed by Xie and Castro (2008). The method allows prescribing the integral length scales and the Reynolds stress tensor. It also allows for the use of spatially varying turbulence scales. The increase in efficiency with respect to the method of Klein et al. (2003) is mainly due to the fact that only one set of 2D random data, rather than a set of 3D data, is filtered to generate a set of 2D data with the prescribed spatial correlations at each time step.

An exponential function with two weight factors is used to correlate the data at the current time step with the one at the previous time step. The other methods use a full 3D digital filter, which is computationally much more expensive. The main validation test cases were channel flow and flow over an array of staggered cubes.

A useful description of the digital filter method can also be found in the Appendix A of Touber and Sandham (2009).

2D vortex method

The random 2D vortex method of Mathey et al. (2003, 2006) deserves attention too because of its simplicity and success in some recent hybrid RANS-LES simulations and because it is already implemented in some commercial CFD codes. This method is mainly suited for generating inflow turbulence at the transition face from RANS to LES zones in hybrid calculations. With this method, velocity fluctuations are generated from a fluctuating two-dimensional vorticity field in the inflow plane perpendicular to the streamwise direction with unit vector $\vec{e}_x$ in this direction. For this, N individual vortices are placed randomly over the inflow plane (with positions $\vec{x}_i = (x, y_i, z_i)$) and the vorticity of those is determined by their circulation Γ_i and a spatial decay function $\eta(\vec{x})$ involving the size of the vortices σ. By superposition of the N individual vortices the fluctuation velocity $\vec{u}'$ at a point $\vec{x}$ in the inflow plane is then calculated from the vorticity distribution of the vortices via the Bio-Savart law as:

$$\vec{u}(\vec{x}) = \frac{1}{2\pi} \sum_{i=1}^{N} \Gamma_i \frac{(\vec{x}_i - \vec{x}) \times \vec{e}_x}{|\vec{x} - \vec{x}_i|^2} (1 - e^{-\frac{|\vec{x} - \vec{x}_i|^2}{2\sigma^2}}) e^{-\frac{|\vec{x} - \vec{x}_i|^2}{2\sigma^2}} \tag{6.5}$$

The fluctuating velocity components v' and w' can be obtained as components of the fluctuating velocity vector $\vec{u}'$. The circulation Γ_i and the vortex size σ appearing in (6.5) are determined from the mean turbulent kinetic energy k and dissipation rate ε via

$$\Gamma_i \propto \sqrt{k(\vec{x}_i)}, \ \sigma \propto k^{\frac{3}{2}}/\varepsilon \tag{6.6}$$

with the distribution of k and ε over the inflow plane known from an upstream RANS calculation (or by estimation). To ensure that the vortices always belong to resolved scales, the minimum value of σ is bounded by the local grid size so that $\sigma \geq \Delta$. The individual vortices are moved randomly to another position at each time step. Further, after a characteristic time scale τ, which is the time the vortices would take to be convected a certain distance in the streamwise direction, the sign of the circulation Γ_i of each vortex is changed randomly.

The vortex method described so far generates only velocity fluctuations v' and w' in the inflow plane normal to the streamwise direction. For generating the streamwise fluctuations u' Mathey et al (2006) suggest a Linear Kinematic Model (LKM) which mimics the effect of the 2D vortices on the streamwise velocity field. The entire procedure of the 2D vortex method is summarized in the flow chart of Figure 6.5.

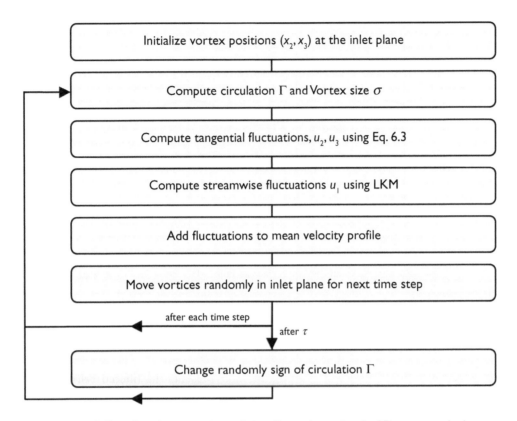

Figure 6.5 Flow chart for generating turbulent fluctuations using the 2D vortex method.

The 2D vortex method was validated successfully for channel and pipe flow and separated flow over periodic hills (Mathey et al., 2003, 2006) and the flow over the Ahmed car body (Mathey and Cokljat, 2005).

6.4 FREE SURFACE BOUNDARY CONDITIONS

The water surface in hydraulic engineering flows represents the boundary between water and the air above it. The water surface can distort into various shapes and will adjust itself according to the flow, turbulence and bathymetric or geometric conditions in the channel. Generally the distortion by turbulence is fairly small and in most cases much smaller than mean surface variations in non-uniform channels, flood waves, tidal channel, ocean waves or flows over hydraulic structures.

Mathematically, at a free surface a *kinematic* as well as a *dynamic* boundary condition applies. *The kinematic boundary condition* states that there is no convective mass transfer through the free surface; hence the fluid velocity component normal to the free surface is equal to the free surface velocity (Fig. 6.6). The kinematic boundary condition in Eulerian form reads:

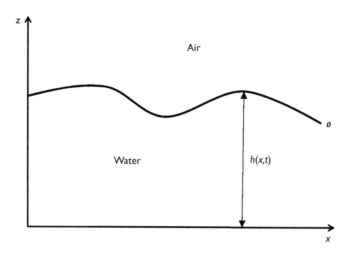

Figure 6.6 Illustration of the free surface as an interface between air and water.

$$\frac{\partial \overline{h}}{\partial t} - \overline{u}_3 = -\overline{u}_1 \frac{\partial \overline{h}}{\partial x} - \overline{u}_2 \frac{\partial \overline{h}}{\partial y} \qquad (6.7)$$

in which $\overline{h}$ is the filtered water depth, and $\overline{u}_i$ represents the filtered velocity components at the surface. The presence of non-linear terms in the unfiltered kinematic boundary condition leads to subgrid scale terms after filtering, which are additional unknown quantities that would need to be modeled but are usually neglected (see e.g. Hodges and Street, 1999). *The dynamic boundary condition* requires that all forces acting at a sharp interface are in equilibrium. In other words, the forces exerted by water on air are equal and opposite to the forces exerted by air on water. Force intensities per unit area that are found at the free surface of an open channel are pressure, acting normal to the surface, viscous stresses, acting both normal and tangential to the surface, and surface-tension, acting both normal and tangential to the free surface. In flows of hydraulic interest, surface tension and viscous stresses can be neglected and the pressure on the air side assumed constant (and is usually set to zero) so that the dynamic boundary condition does not apply.

The implementation of the kinematic boundary condition into the solution algorithm of the filtered Navier Stokes equations is not straightforward because neither shape nor position of the free surface are known. There are two principal numerical methods for calculating the location of the free surface (Ferziger and Peric, 2002), i.e. *Interface-Tracking Methods* (ITM) and *Interface-Capturing Methods* (ICM), also known as moving-mesh and fixed-mesh methods. In an interface-tracking, or moving-mesh, method for open-channel flow, the governing equations are only solved for the water phase and the computational mesh is moved after every time step so that the boundary of the computational domain coincides with the free surface geometry. An ITM requires a boundary-fitted, moving grid together with an appropriate algorithm that readjusts the grid in the entire domain each time the surface has moved. So far,

ITMs have been used mainly in the context of RANS modelling, following different strategies to compute the free-surface elevation. An ITM in the context of LES has been presented by Hodges and Street (1999) who simulated the interaction of waves with a turbulent channel flow. These authors used an explicit time-discretization scheme to advance the free surface by solving the kinematic boundary condition and solved a Poison-type equation after every time step to compute a new boundary-orthogonal grid (the surface and the mesh on the surface at one instant in time are shown in Figure 6.7). The Reynolds number in this case is rather low (Re_{τ} = 171) so that the turbulent eddies and the surface deformations caused by them have rather large length and time scales. At Reynolds numbers of practical interest with much smaller turbulent length and time scales the recalculation of a new mesh would be extremely expensive. In fact, Hodges and Street state that in such cases of small surface deformations their method is not suitable. To avoid the creation of a new mesh after every time step Fulgosi et al. (2003) used a mapping scheme that transfers the curvilinear physical space into an orthogonal coordinate system. Fulgosi et al. employed this technique for a DNS of wind-sheared free-surface deformations.

In *interface capturing methods* the boundary between the two phases is not defined sharply through the numerical mesh, hence the mesh includes both phases (Fig. 6.8). Probably the earliest ICM was proposed by Harlow and Welch (1965). Their Marker-and-Cell (MAC) method introduces massless particles into the water near the free surface, which are moved according to the velocity components in their vicinity. Though the MAC scheme can handle complex interfaces such as breaking waves, a large number of particles are needed, which makes the method computationally expensive. Alternatively the free surface position can be determined by the contour of a scalar function which does not need to coincide with grid lines. The Volume of Fluid method (VOF) is such an approach, where the fraction of the liquid phase is

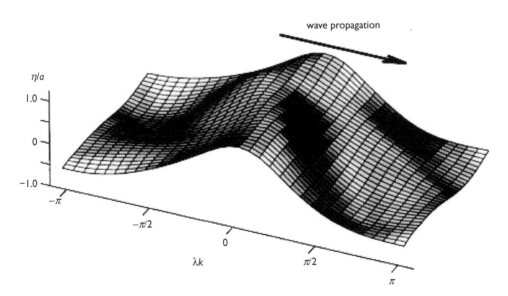

Figure 6.7 Mesh deformation for the simulation of wave interaction with a turbulent flow using an Interface Tracking Method (from Hodges and Street, 1999).

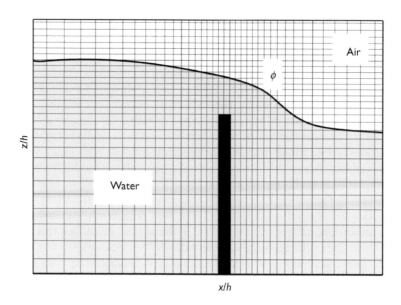

Figure 6.8 Location of the free surface in the flow over a sharp crested weir using an Interface Capturing Method. The computational domain contains both water and air phase.

determined by the solution of a transport equation for the void fraction F (Hirt and Nichols, 1981). Per definition F is unity in any cell/point that is occupied by water and zero otherwise. One important procedure in VOF methods is that the surface has to be reconstructed in terms of the volume fraction for which different techniques exist (e.g. SLIC, Simple Line Interface Calculation method by Noh and Woodward, 1976, or PLIC, Piecewise Linear Interface Calculation, by Youngs, 1982). Thomas et al. (1995) combined the kinematic boundary condition based on the water depth h (Eq. 6.4) with the VOF method, which is applied to a LES of straight open channel flow by Shi et al. (2001). Recently, the Level-Set Method (LSM), which originated in computer graphics has become a popular interface-capturing method for multiphase flows. The LSM was originally proposed by Osher and Sethian (1988) and was developed for computing and analyzing the motion of an interface between two phases in two or three dimensions. In the LSM the conservation equations are solved for both liquid and gas phase and the interface moves at the local velocity which can be expressed in a Lagrangian way. The LSM has been recently employed in an LES of the flow over fixed dunes by Yue et al. (2005). Not only were Yue et al. (2005) able to accurately and realistically calculate the unsteady free surface motion but they also provided evidence of boils, upwelling and downdraft.

By far the most large-eddy simulations in hydraulics were performed with the so-called rigid-lid approximation as boundary condition for the free surface. In this, instead of the actual free surface as adjusting boundary in an interface-tracking method, a fixed boundary, generally a flat surface, is used which is treated as a frictionless wall. Hence, the calculations are in fact carried out for a closed conduit with an artificially introduced fixed upper boundary where the shear stress is zero (in the absence of wind shear) and the velocity normal to the boundary is also zero, but where the pressure can vary as it does along a wall. In fact, the boundary conditions for the pressure and the normal velocity are as described earlier for a wall, while the shear stress at the boundary is simply set to zero which altogether amounts to using symmetry conditions at the boundary. The problem of having to determine the surface elevation while knowing the pressure at the surface has been shifted to knowing the location of the boundary but having to determine the variation of the pressure there. The surface-elevation-gradient terms $g\partial h/\partial x$ and $g\partial h/\partial y$ in the momentum equations for free surface flows are thereby replaced by the pressure gradients $\partial p/\partial x$ and $\partial p/\partial y$ so that the dynamic effects of surface-elevation variations are properly accounted for by the rigid lid approximation method. In fact, the surface elevation that would evolve can be retrieved from the pressure variations through the following relation:

$$\Delta h_\kappa = z_\kappa - z_{ref} = \frac{p_\kappa - p_{ref}}{\rho g} \tag{6.8}$$

where p_κ and z_κ are local pressure value and surface elevation and p_{ref} and z_{ref} are the corresponding quantities at a reference location. However, by suppressing the actual surface deformation a certain error is introduced in the continuity equation, but this is small when the surface deviation is small compared with the local water depth, say below 10% of the depth. This is virtually always the case for surface disturbances due to turbulence and hence the rigid-lid approximation is the preferred free-surface boundary condition in DNS and LES calculations. For example, Lam and Banerjee (1992) and Pan and Banerjee (1996) studied the turbulent structures in DNS of open channel flow using the rigid-lid approximation. Komori et al (1993) included the surface variations in their computation by solving Equation (6.7) for the water depth and compared the results with the ones from Lam and Banerjee (1992). They found that the free-surface deformations and the normal velocity at the free surface remained extremely small so that the calculated flow behaviour near the free surface and in particular the fluctuations there did not differ from the simulations using a flat rigid lid. The rigid-lid approximation was also used in many RANS calculations and in LES for situations with relatively small mean surface variations. However, larger errors must be expected for the mean variations when these are not small compared with the local water depth as can be the case in coastal and ocean engineering problems where the variation of the mean free surface is usually much more pronounced. A wide variety of applications of the rigid-lid approximations in LES of hydraulic flow situations will be presented in the Applications Chapter 9. In fact, all the LES presented and discussed there for free-surface situations were obtained with this approximation.

6.5   SMOOTH-WALL BOUNDARY CONDITIONS

The no-slip condition at walls is also valid for turbulent flow and its fluctuating velocities, and hence the natural boundary condition there is for impermeable walls to set all components of the resolved velocity to zero, i.e. $\overline{u}_{i\,w} = 0$. Because of steep gradients and strongly reduced size of the dominant eddies very near walls, high numerical resolution is necessary, which means that several grid points must be placed within the viscous sublayer. When a finite-volume method is used, it is not the velocity at the wall that must be specified but the momentum flux at the lower boundary of the wall-adjacent cell, which is the wall shear stress (see Fig. 6.9). The wall shear stress is equal to the molecular viscosity times the velocity gradient at the wall, i.e.

$$\tau_w = \mu \left(\frac{\partial \overline{u}_t}{\partial z} \right)_w \tag{6.9}$$

in which $\overline{u}_t$ is the resolved velocity parallel to the wall. In the viscous sublayer the velocity increases linearly with distance from the wall, and since this sublayer has to be resolved properly in this approach, the gradient in (6.9) can be replaced by $\overline{u}_{t_1} / z_1$ where z_1 is the distance of the first grid point from the wall which is located in the middle of the wall-adjacent cell (see Fig. 6.9). This allows calculating the instantaneous wall shear stress from the resolved tangential velocity at the first grid point. Concerning the boundary condition for the pressure, there are several possibilities which depend on the flow problem considered and also on the numerical procedure used. In fully-developed straight channel flow without obstacles, the exact pressure condition is a zero wall-normal gradient von Neumann condition, i.e. $\frac{\partial p}{\partial z}\big|_w = 0$. In the more general case with curved walls, e.g. the flow around an obstacle or in a bend, this condition does not hold because wall-normal pressure gradients occur. In this case the pressure at the boundary is approximated through a linear extrapolation from the known values inside the flow domain. When the finite-volume procedure for solving the equations uses a staggered grid, pressure boundary conditions need not be specified at all.

In the region very close to the wall (viscous sublayer, buffer region) the major part of the turbulence production occurs and small-scale structures such as high- and low-speed streaks must be resolved. An approach achieving this is called a wall-resolving LES. For this, a good resolution is necessary not only normal to the wall but also in the wall-parallel directions, and in fact the resolution must be similar to that in a DNS. Hence this approach is called a Quasi-Direct Numerical Solution (QDNS, Spalart et al 1997). Chapman (1979) was the first to analyze the resolution requirements of flow over a flat plate and found that $z_1^+ \approx 1$, $\Delta x^+ \approx 100$, $\Delta y^+ \approx 20$ is necessary as minimum resolution. This was later confirmed by other authors, e.g. Piomelli and Balaras (2002). Chapman (1979) also estimated that the number of grid points required for such an LES is proportional to $Re_L^{1.8}$. For channel flow Bagget et al (1997) estimate the required grid points to be proportional to Re_τ^2, where Re_τ is the friction Reynolds number (see 9.1).

It is clear that at high Reynolds numbers of practical interest, calculations with such fine resolution are not feasible. Such LES are cheaper than a complete DNS but they are still too expensive. A way out is to avoid resolving the details of the structures in the near-wall region by LES through the use of special near-wall models. One approach is to place the first grid point outside the region where resolution is particularly expensive, generally in the log-law region (Fig. 6.9b), and to bridge the near-wall region by wall functions. Another approach is to calculate the near-wall region by a RANS model, which also requires quite fine resolution in the wall-normal direction but not in the wall-parallel directions so that considerable saving is achieved. This

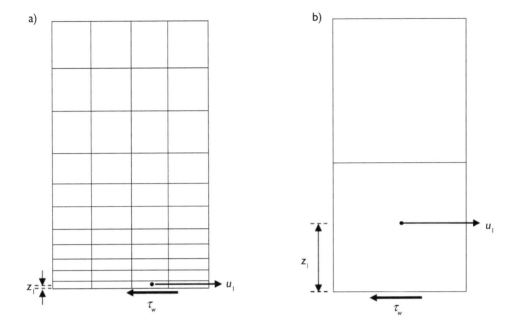

Figure 6.9 Near-wall numerical mesh cells for finite-volume LES.
a) for wall resolving LES.
b) for LES with wall function.

approach will be dealt with in Chapter 7 on Hybrid RANS-LES methods. In both cases, the near-wall models must provide the wall shear stress but also proper information on the fluctuations for the resolved region further away from the wall. Hence such models can be considered as a kind of special subgrid-scale model.

In the following, the most commonly used wall functions are introduced; these basically relate the instantaneous wall shear stress to the resolved velocity at the grid point placed in the middle of the wall-adjacent cell in a finite-volume procedure (see Fig. 6.9).

Schumann's model

The most popular wall-function model is due to Schumann (1975) who proposed the following relation

$$\bar{\tau}_w = \frac{\langle \bar{\tau}_w \rangle}{\langle \bar{\tau}_1 \rangle} \bar{u}_1 \tag{6.10}$$

between the instantaneous wall shear stress $\bar{\tau}_w$ and the instantaneous velocity $\bar{u}_1$ at the first grid point, assuming that these are in phase and that the proportionality factor is the ratio of time-averaged shear stress and velocity at the first grid point,

$< \bar{\tau}_w > / < \bar{u}_1 >$. Schumann (1975) further assumed that this ratio is determined by the following logarithmic law of the wall

$$<\bar{u}^+> = \frac{<\bar{u}>}{u_*} = \frac{1}{\kappa}\ln(z^+) + B \qquad (6.11)$$

i.e. that the first grid point lies in a region where this law prevails. In (6.11), $z^+ = zu_*/v$ is the normal distance from the wall in wall units, u_* is the friction velocity $u_* = \sqrt{<\bar{\tau}_w>/\rho}$, κ is the von Karman constant (= 0.4) and B another constant depending on the roughness of the wall ($\approx$ 5.0 for smooth walls). The log law (6.11) is valid in the z^+-range $30 \le z^+ \le 500$.

The time-averaged values $<\bar{u}_1>$ and $<\bar{\tau}_w>$ have to be determined by time-averaging during the time-marching solution; for flows with homogenous directions this can be supported by averaging in these directions. Further, an iterative solution of (6.11) is necessary to determine the friction velocity u_* and hence the wall shear stress $<\bar{\tau}_w>$ from the velocity $<\bar{u}_1>$. In the special case of developed channel flow, the time-averaged wall-shear $<\bar{\tau}_w>$ is known from the imposed pressure gradient.

For practical applications it is important that the wall-function relation (6.10) can be applied also when the first grid point comes to lie in the viscous or buffer layer. This can be achieved by replacing (6.11) with the following three-layer formulation for the velocity distribution (layer I: viscous sublayer, layer II: buffer layer, layer III: logarithmic layer):

$$<u^+> = z^+ \qquad I: \quad \text{for } 0 \le z^+ < 5 \qquad (6.12a)$$
$$<u^+> = a_1 \ln(z^+) + b_1 \quad II: \quad \text{for } 5 \le z^+ < 30 \qquad (6.12b)$$
$$<u^+> = a_2 \ln(z^+) + b_2 \quad III: \quad \text{for } 30 \le z^+ < 300 \qquad (6.12c)$$

with constants $a_1 = 5.0$, $a_2 = 2$ or $1/\kappa$, $b_1 = -3.05$, $b_2 = 5.2$.

In complex flows, the direction of the wall shear stress is not known a priori and the 2 components $\bar{\tau}_{w,x}$ and $\bar{\tau}_{w,y}$ have to be determined from the 2 wall-parallel components of the velocity at the first grid point, $\bar{u}_1$ and $\bar{v}_1$. This is achieved with the following relations:

$$\bar{\tau}_{w,x} = \frac{<\bar{\tau}_w>}{\sqrt{<\bar{u}_1>^2 + <\bar{v}_1>^2}}\bar{u}_1, \quad \bar{\tau}_{w,y} = \frac{<\bar{\tau}_w>}{\sqrt{<\bar{u}_1>^2 + <\bar{v}_1>^2}}\bar{v}_1 \qquad (6.13)$$

Werner-Wengle model

The Schumann model is based on relations derived from flat-wall situations and requires the determination of mean values and an iterative procedure for obtaining the mean wall shear stress from the logarithmic velocity law. In order to avoid the latter possibly elaborate procedures and aiming at an application to more complex situations, Werner and Wengle (1993) proposed a wall-function model that does not involve a law for the time-averaged values but directly a distribution for the instantaneous velocity, assuming again that the instantaneous wall shear stress is in phase with this. A two-layer distribution is assumed, with a linear profile for the inner, mainly viscous layer and a power law for the outer turbulent layer:

$$< u^+ > = z^+ \qquad I: \quad \text{for } 0 \leq z^+ < 11.82 \tag{6.14a}$$

$$\bar{u}^+ = a_2 (z^+)^m \qquad II: \quad \text{for } 11.82 \leq z^+ < 1000 \tag{6.14b}$$

with the parameters $a_2 = 8.3$ and $m = 1/7$. The change-over from the inner to the outer layer is at $z_m^+ = 11.8$. Integration of these velocity distributions over the wall-adjacent cell in a finite-volume procedure then yields directly the wall shear stress:

$$\bar{\tau}_w = \rho \frac{2\nu}{\Delta z_1} \bar{u}_1 \qquad \bar{u}_1 \leq \frac{\nu}{2\Delta z_1}(z_m^+)^2 \tag{6.15a}$$

$$\bar{\tau}_w = \rho \left[\frac{1-m}{2} C_m^{\frac{1+m}{1-m}} \left(\frac{\nu}{\Delta z_1} \right)^{1+m} + \frac{1+m}{C_m} \left(\frac{\nu}{\Delta z_1} \right)^m \bar{u}_1 \right]^{\frac{2}{1+m}} \qquad \bar{u}_1 > \frac{\nu}{2\Delta z_1}(z_m^+)^2 \tag{6.15b}$$

where $\Delta z_1 = 2z_1$ is the height of the wall-adjacent cell (see Fig. 6.9) and $C_m = 8.3$. In complex flows, the wall-shear stress components can then be determined from

$$\bar{\tau}_{w,x} = \frac{\bar{\tau}_w}{\sqrt{\bar{u}_1^2 + \bar{v}_1^2}} \bar{u}_1 \qquad \bar{\tau}_{w,z} = \frac{\bar{\tau}_w}{\sqrt{\bar{u}_1^2 + \bar{v}_1^2}} \bar{v}_1 \tag{6.16}$$

The Werner-Wengle model is applicable to situations with flow separation and has been used with some success for such flows by the proposers and other authors, even though the assumed velocity distributions are questionable in these regions.

Piomelli et al's model

Piomelli et al (1989) extended Schumann's model based on the experimental observation that the correlation between wall shear stress and velocity at a near wall point increases when there is a time delay between velocity and shear stress. This is due to the also experimentally observed fact that the near-wall structures mainly responsible for the velocity and shear-stress fluctuations are inclined. Piomelli et al (1989) accounted for this in their model by relating the wall shear stress in point A to the velocity at a point B along an inclined near-wall structure (see Fig. 6.10). Hence they proposed to replace (6.10) by the "shifted boundary condition":

$$\bar{\tau}_w = \frac{<\bar{\tau}_w>}{<\bar{u}_1>} \bar{u}_1 (x + \Delta s) \tag{6.17}$$

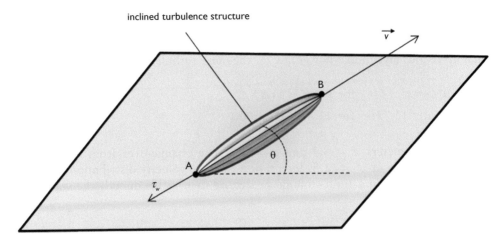

Figure 6.10 Sketch illustrating Piomelli et al's (1989) wall function based on an inclined near-wall structure.

In which Δs is the streamwise displacement of point B with respect to point A and is defined via the angle of inclination of the structures, Θ, and the wall distance of point B, y_1. From experiments and DNS the optimal inclination angle was determined to lie in the range $\Theta = 8° - 13°$. For $30 < y_1^+ < 50$ the smaller value applies so that $\Delta s = z_1 \cot 8°$ while for larger wall distances $\Delta s = z_1 \cot 13°$ yields better results. For plane channel flow, the inclusion of the displacement Δs improved the results over Schumann's formulation, and for this flow Balares et al (1995) obtained excellent agreement with experimental and DNS data for a range of Reynolds numbers. For more complex flows, where the mean-flow direction is not known a priori, the direction of the displacement must also be determined, which complicates the application of the model.

6.6 ROUGH-WALL BOUNDARY CONDITIONS

The above wall boundary conditions/models are applicable to geometrically simple flows over smooth walls for which the near-wall flow can be resolved or approximated with some model. However, in flows of hydraulic-engineering interest walls, e.g. the bed of a river, are often rough so that one of the challenges of LES of such flows is the accurate inclusion of wall roughness. Similar to flow over smooth walls, specification of the wall shear stress or some a-priori knowledge of the velocity profile over the rough wall is needed. This by itself is not an easy task as the law of the wall for rough boundaries is still subject of ongoing research and several different formulations exist. In his overview, Patel (1998) therefore calls the implementation of rough wall boundary conditions the "Achilles Heel of CFD", even though Patel's article aimed at the RANS modelling community. There are several different concepts of dealing with rough walls in LES. The most common ones are introduced in the following.

Explicit resolution of roughness elements

One basic possibility to account for wall roughness is to resolve the flow around the individual roughness elements, specifying as boundary condition the no-slip condition (zero velocity) at the walls of these elements.

The success of Direct Numerical Simulations (DNS) and Large Eddy Simulations (LES) in revealing details of the turbulent channel flow over smooth walls has initiated DNS and LES studies of flow over walls with roughness elements. Such fundamental studies resolved the individual elements through the numerical grid, the numerical effort thereby being extremely high. Successful LES of that kind are reported for flows over relatively simple and exactly defined roughness elements like square bars (e.g. Leonardi et al., 2003; Ikeda and Durbin, 2007, Stoesser and Nikora, 2008), wavy walls (e.g. Calhoun and Street, 2001, Nakayama et al., 2004) spheres (Singh et al., 2007) or typical open-channel bedforms such as sand dunes or ripples (e.g. Yue et al., 2005 a, b, 2006; Stoesser et al., 2008; Zedler and Street, 2001). An alternative to using body-fitted grids to explicitly resolve individual roughness elements is the Immersed Boundary Method (IBM) as employed for instance by Lee (2002), Cui et al. (2003), and Bhaganagar et al. (2004), and which is described in Chapter 4. In high-Re flows of hydraulic interest, e.g. open-channel flow over rough beds, explicit resolution of the flow around roughness elements has had limited success mainly for two reasons: (1) the detailed topology of the rough surface is unknown or (2) the requirements in terms of grid resolutions to represent individual roughness elements (using ten or more points per roughness element) are exceeding current computing resources. For instance, a LES (employing the IBM) of the flow over a rough surface that is composed of small sandgrains (e.g. $H/k = 30$) in a $6H \times 3H \times H$ domain requires approximately 0.2×10^9 gridpoints, with each sandgrain being resolved by only 10^3 points. Moreover, in open-channel flow over an alluvial bed the rough surface comprises sediments of widely different size which can range from micro-meters to decimeters with relative submergences ranging from one to several hundred, making the specification of accurate boundary conditions an almost insurmountable challenge. Hence, for calculating flows of practical interest, alternative methods must be used in which the individual roughness elements are not resolved.

Rough-wall law based boundary conditions

To date most numerical simulations of flow over rough walls have been based on the Reynolds-Averaged Navier Stokes (RANS) equations, where the effect of roughness has been accounted for by wall functions involving roughness functions determined empirically from experiments. This RANS-type treatment is, in theory, applicable in LES too, in fact it is quite common in LES of atmospheric-boundary layer flow. The first grid point is placed outside the roughness layer (see Figure 6.11), which is the (rough) equivalent of the buffer layer in smooth walls, and an instantaneous shear stress is imposed at the boundary, in effect at the lower boundary of the first grid cell. Following the approach by Schumann (1973) discussed above, this stress can be determined from Equation (6.10), with the ratio of time-averaged shear stress and velocity at the first grid point, $<\overline{\tau_w}>/<u_1>$ now obtained from a logarithmic law of the wall for rough walls:

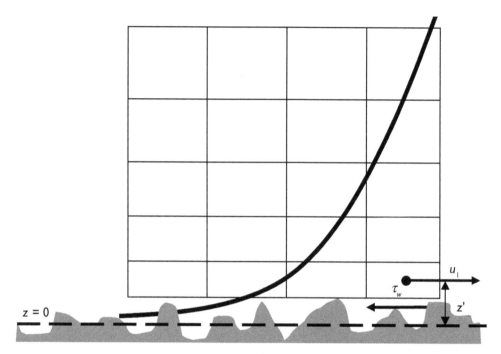

Figure 6.11 Near-wall numerical mesh cells for finite-volume LES over a rough boundary.

$$\frac{<u>}{u_\circ} = \frac{1}{\kappa}\ln(z^+) + 5.5 - \Delta B \tag{6.18}$$

The last term on the RHS of Equation (6.18), ΔB, expresses the downshift of the velocity profile (from the one over a smooth wall) when plotted on a semi-logarithmic Clauser plot. The magnitude of the downshift ΔB depends on the equivalent grain roughness, k_s, which is a common length scale used to quantify the roughness. The roughness Reynolds number, $k_s^+ = u_\circ k_s/\nu$ is used to distinguish between the three roughness regimes, hydraulically smooth, transitional roughness, and fully rough for which ΔB is calculated as (Cebeci and Bradshaw, 1977):

$$\Delta B = \begin{cases} 0, & \text{for } k_s^+ < 2.5 \\ \left[5.5 - 8.5 + \dfrac{1}{\kappa}\ln(k_s^+)\right]\sin\left[0.426(\ln k_s^+ - 0.81)\right] & \text{for } 2.5 \le k_s^+ < 90 \\ 5.5 - 8.5 + \dfrac{1}{\kappa}\ln(k_s^+) & \text{for } k_s^+ \ge 90 \end{cases} \tag{6.19}$$

In hydraulics there are different types of bed roughness (van Rijn, 2007), the two most relevant are grain roughness and bed-form roughness. For smooth beds $k_s \approx 0$, and the smooth bed log law applies. For laboratory channels in which a smooth bed is technically roughened with individual sand grains the equivalent grain roughness is approximately the mean grain diameter, i.e. $k_s \approx d_{50}$. In natural beds, or in laboratory channels with a gravelly or a sandy bed, higher sandgrain roughness values are recommended because larger grains will dominate the surface as a result of bed armoring or washing out of finer grains. Therefore, van Rijn (1984) recommended $k_s = 3d_{90}$, however other approximations are available depending on the bed material e.g. Einstein (1942) recommended $k_s = d_{65}$ for sand; Hey (1978) suggested $k_s = 3.5d_{84}$ for coarse gravel.

For bed forms such as ripples or dunes, both grain roughness and form roughness have to be accounted for. A common approximation is given by van Rijn (1984) as $k_s = 3d_{90} + 1.1\Delta\left(1 - e^{-25\Delta/\lambda}\right)$, in which Δ is the bed-form height and λ is the bed form length.

Once the equivalent grain roughness has been determined, the shear stress at the lower boundary of the first grid cell above the virtual origin (i.e. $z = 0$), $\overline{\tau_w}$, can be calculated as a function of the filtered horizontal velocity ($\overline{u}_i$) at the first grid point, which is located some vertical distance, z', above the rough boundary (see Figure 6.11). The components of this shear stress can be expressed through the following quadratic relation as:

$$\tau_{w_i} = - < \overline{\tau_w} > \frac{\overline{u}_i}{< u(z') >} = c_D \left[< u(z') >\right]^2 \frac{\overline{u}_i}{< u(z') >} \tag{6.20}$$

where $\overline{u}_i$ are the filtered, tangential velocity components at the first grid point off the virtual rough wall and $< u(z') >$ is the time-averaged streamwise velocity at this location. c_D is a drag coefficient, which, in case of a fully rough boundary (i.e. for for $k_s^+ \geq 90$ in Eq. 6.19), is determined by:

$$c_D = \kappa^2 \left[\ln(30z / k_s)\right]^{-2} \tag{6.21}$$

Equation (6.21) can be considered the hydraulic-engineering analogue to what has been used widely in the atmospheric boundary layer community. There, the formulation of the drag coefficient is as follows (Moeng, 1984, Porte-Agel et al., 2000, Chow et al, 2005):

$$c_D = \kappa^2 \left[\ln(z / z_0)\right]^{-2} \tag{6.22}$$

where z_0 is a roughness length that has to be determined from experiments. When comparing Equation (6.22) with Equation (6.21) it is obvious that $k_s = 30z_0$.

Instead of obtaining the instantaneous wall shear stress through the time-averaged velocity, Saito et al. (2012) derived a dynamic ordinary differential equation, which allows calculating the instantaneous friction velocity, u_*, at the virtual origin. Once u_* is obtained, an instantaneous version of equation (6.18) can be solved explicitly for an instantaneous local slip velocity at the virtual origin.

The validity and accuracy of a log-law based boundary condition in LES of channel flow over rough beds is currently a topic of ongoing research. There are several uncertainties and assumptions involved in this treatment. For instance, a relatively coarse grid near the roughness may lead to inaccuracies of the velocity gradient or for flows with low water-depth-to-roughness-height the grid is too coarse over the entire water depth to resolve the dominating turbulent structures. Another uncertainty includes the virtual origin of the rough bed, which can result in inaccuracies regarding the shear stress or for large roughness in a continuity defect. However, one of the greatest uncertainties of using wall functions is the fact that due to the absence of local variations in bathymetry, the effects of local pressure gradients and streamline curvature on the flow are neglected. In particular if the bed is comprised of exposed roughness elements, local flow separation and recirculation can be substantial contributors to turbulence production and are likely to affect the flow over a substantial portion of the water depth. However, for flows at high relative submergence the log-law based semi-slip condition can provide reasonable results if (1) the effect of the roughness on the velocity profile can be estimated *a-priori* (e.g. by knowledge of the equivalent sand-grain roughness and zero-plane displacement) (2) the interaction of the flow in the roughness layer with the outer layer is very weak or absent, which is true if the roughness height is small. For this to hold Jimenez (2004) proposes water-depth-to-roughness-height >> 40.

Momentum forcing

A more natural way to account for the turbulence producing roughness but also avoiding the expensive explicit resolution of roughness is to consider the rough bed in a spatially averaged sense. In the momentum forcing approach the effect of the roughness on the flow and turbulence above the rough bed over a predefined volume is simulated. Such an approach was presented by Nakayama and Sakio (2002) who suggested adding extra dispersive stress terms to the momentum equations and using a slip boundary condition at a suitably chosen boundary to mimic the roughness. The dispersive stresses are a result of spatially averaging the Navier-Stokes equations over a control volume in which the velocity profiles are heterogeneous in space due to the rough bed (Nikora et al., 2006). The method of Nakayama and Sakio (2002) reproduces the flow statistics fairly well but requires a priori knowledge of the magnitude of the dispersive stresses, which they obtained from a DNS simulation.

Also for vegetation roughness; to avoid the computationally very expensive resolution of vegetation elements, less costly approaches were developed in the meteorological community using a vegetative momentum force. The idea is to compute a drag force and add it to the right hand side of the momentum equations such as the following time-dependent drag force for vegetation (e.g. Shaw and Shumann, 1992 or Kanda and Hino, 1994):

$$F = C_D a V^2 \frac{\overline{u_i}}{|V|} \tag{6.23}$$

in which C_D is a drag coefficient, a is the leaf density in [m²/m³], and $|V|$ is the magnitude of the velocity vector ($\sqrt{\overline{u_i u_i}}$). The drag force acts as a momentum sink and causes energy losses by stem and leaf drag. However, this treatment alone does not produce realistically the turbulence resulting from the flow around the plants and the wake flow downstream of the plants. Hence, a modification of the SGS stresses is also needed. In meteorological canopy models, the SGS viscosity is mostly calculated using a one-equation subgrid-scale model in which SGS energy loss due to plant drag as well as SGS wake production are accounted for by additional drag-related source terms. Probably the greatest weakness of such LES of canopy flow is that the drag needs to be parameterized through a drag coefficient that is generally not known a priori. Values between 0.1 and 1.0 were used and are based on measurements. For instance, Shaw and Schumann (1992) used $C_D = 0.15$ for the flow within and above a forest, while Kanda and Hino (1994) used $C_D = 0.5$ for flow within a plant canopy layer. In an LES of open-channel flow over and through submerged vegetation, Cui and Neary (2002) also employed momentum forcing and used $C_D = 1.0$. Further empirical parameters are needed in the subgrid-scale models of flow through vegetation/plant canopies. The interested reader is referred to Shaw and Schumann, 1992, Kanda and Hino, 1994, or Dwyer et al., 1997.

In order to avoid the use of empirical drag coefficients for the near-wall treatment, Scotti (2006) proposed to mimic roughness within the framework of DNS/LES. The idea is to randomly place ellipsoids on a smooth wall (Figure 6.12a) aiming at imitating sandpaper roughness. A Cartesian grid is used overlying the roughness and the grid spacing is determined from the resolution requirements for an LES/DNS and not based on the size of the roughness elements, which can be larger or smaller than the grid spacing. Scotti's method is as follows: (1) place ellipsoids of height k and length $2k$ on the flat bed and calculate the solid and fluid volume fraction in each grid cell (2) solve the momentum equations to obtain an intermediate velocity (3) correct the intermediate velocity by multiplying it with the fluid volume fraction (4) solve for pressure and correct velocity to ensure divergence-free flow field. This treatment reproduces the downward shift of the spatially averaged velocity profile on a Clauser plot as well as realistic spatially averaged turbulence statistics. Stoesser (2010) extended Scotti's method from technical roughness to natural roughness by generating a realistic, natural-channel bed topology from measurable physical parameters. Stoesser's (2010) method requires knowledge of the mean grain diameter d_{50} of the sediment material comprising the rough bed. The material is then placed on the bed according to a roughness-geometry function of the variation in bed elevation z' from the mean z_{bm} (Figure 6.12b), which was proposed by Nikora et al. (2001) for alluvial channels and later confirmed by Aberle and Nikora (2006) for a gravel bed in a laboratory flume. A Cartesian grid is then generated, which in part includes the rough bed (see Figure 6.12c) and momentum forcing is applied to force the velocity to zero in the cells that contain the roughness. Stoesser's method is as follows: (1) generate rough-bed topography using d_{50} and roughness-geometry function involving a random number generator. (2) solve the momentum equations including a momentum force to attain zero velocity in computational cells occupied by the roughness (3) solve for pressure and correct velocity to ensure divergence free flow field. Stoesser (2010) carried out several LES of flow over sand and gravel beds to validate the method and spatially averaged turbulence statistics of the simulations matched measured values fairly well (see Figure 9.7 in section 9.2).

The advantage of both the Scotti and the Stoesser methods is that they are extremely simple to implement numerically. Also they only require physically measurable input

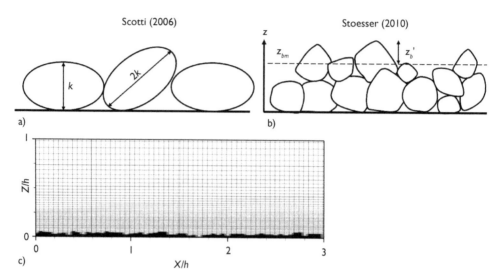

Figure 6.12 Representation of bed roughness within the framework of LES/DNS for technical roughness (a) and natural roughness (b) and artificially generated roughness immersed in
a Cartesian grid (c).

parameters, hence they do not rely on empirical formulae or subjective roughness
length scales. This is especially valuable for flows of hydraulic interest in which the
details of the rough bed are only known approximately or in a statistical sense.

An other interesting momentum forcing approach for flow over rough terrain
was proposed by Anderson and Meneveau (2011). The approach makes use of the
fact that elevations of the rough terrain follow a power law spectrum. In Anderson
and Meneveau's method a synthetic rough terrain surface is constructed by using
random-phase Fourier modes with prescribed power-law spectra and is then decomposed into resolved and sub-grid scale height contributions. An immersed drag force is
used to account for the resolved roughness, similar to Stoesser's method. The *a-priori*
unknown sub-grid roughness contributions are accounted for through a local equilibrium wall model, for which the required roughness parameter is obtained from a
Germano-identity-type test-filtering procedure.

6.7 INITIAL CONDITIONS

The specification of initial conditions for large-eddy simulations requires values of
each variable (velocity, pressure, scalar) in the entire computational domain. The cor-
rectness of these initial values is only important if the calculated results are needed
from the start of the simulation. In large-eddy simulations of hydraulic problems this
is hardly the case; rather flow and turbulence of a fully developed state are of interest.
In analogy to laboratory experiments in hydraulics, LES requires a certain approach
length (in the laboratory it is the channel entrance section while in most LES it is the

initial simulation period) for the flow to develop and to reach a state that is independent of the previously specified initial conditions. In order to estimate the length of the initial simulation period the flow-through time unit T_{ft}, defined as

$$T_{ft} = u(bulk)/(length\ of\ domain) \qquad (6.24)$$

or the eddy turn-over time unit T_{to} defined as:

$$T_{tot} = u_*/(water\ depth), \qquad (6.25)$$

where u_* is the friction velocity, can be employed. The period in terms of these two time units however depends on the flow situation and the Reynolds number, but as a rough guideline, at least six time units of either one are needed for the flow to be considered fully developed. It is recommended to monitor the velocity or pressure signal at various points in the flow to check the validity of the estimates made and make modifications to the initially chosen development period if necessary.

Chapter 7

Hybrid **RANS-LES** methods

7.1 INTRODUCTION

7.1.1 Motivation

As discussed already in Chapter 1, RANS methods can predict the flow quantities of interest to hydraulic engineers with sufficient accuracy for many relatively simple flows, that is when the flow is mainly unidirectional and largely attached as in channel-type flows. They still are the main workhorse of hydraulic engineers and allow practical flow calculations at reasonable cost, and not only at laboratory scale but also in field situations with high Reynolds numbers. However, RANS models have been found to be not sufficiently adequate for complex situations where large-scale anisotropic structures are important and govern the flow behaviour and where unsteady effects are of interest such as unsteady forces on structures and situations with fluid-structure interaction. In such cases LES has been found to be clearly superior, but it is also much more expensive as LES will always be three-dimensional and time-dependent, and near walls the resolution requirements are such that the number of grid points necessary increases roughly with $\text{Re}^{1.8}$ for a truly wall-resolving LES (Chapman 1979). Hence such LES are not affordable for real situations with large Reynolds numbers as they usually occur in practice.

The different characteristics of the two types of methods and their different strengths and weaknesses for different application areas suggest that the resolution and applicability problem could be overcome by combining the two approaches, and here two scenarios have emerged. One is to use LES only where needed because the flow is too complex to be accurately predicted by RANS or where information on the dynamics of the flow is required. In hydraulics this is often the case in only part of the flow domain, e.g. in rivers in the vicinity of structures such as groynes, bridge piers, or other geometric complexities such as confluences, bends, etc. – while in other parts the flow is close to developed channel flow and hence amenable to RANS calculation with sufficient accuracy for practical purposes. Therefore, in such an approach only a sub-region where complex flow prevails is simulated with the expensive LES method while the rest would be calculated with a RANS method. This approach is called embedded LES and will be discussed in some detail below.

The other scenario, which was in fact the main driver for the development of hybrid RANS-LES models, relates to the near-wall problem at high Reynolds numbers mentioned above. In the early days of LES, this near-wall resolution problem was circumvented with the aid of wall functions which bridge the near-wall region that would require high resolution in a wall-resolved LES. With this approach, which is introduced in chapter 6 on boundary conditions, the near-wall region containing complex fine-scale turbulent structures is not resolved. Rather, only the global effect of this region on the outer resolved region is simulated, e.g. by making use of the log law or exponential profile assumptions for the near-wall velocity. However, these wall-functions are not generally valid and not so suitable for geometrically complex situations with separation and near-wall secondary motion, etc. so that their application has not been very successful in such cases. More recently, the near-wall resolution problem was attempted to be overcome by the hybrid approach, namely by simulating the near-wall region with RANS (or rather unsteady RANS called URANS) and only regions further away from walls, where the turbulent eddies are larger, with LES. A well-resolved LES would require high resolution in all directions to resolve the complex near-wall structures. Meanwhile, with RANS the resolution can be considerably lower, particularly in the wall-parallel directions. The RANS resolution does of course not allow obtaining the near-wall structures but aims to provide realistic mean-flow and turbulence statistics. In a sense, such an approach amounts to a refined wall model. This treatment requires the use of a low-Reynolds-number RANS model which also needs a fairly fine resolution, but only in the wall-normal direction. In hydraulics, the walls are usually rough and hence the low-Reynolds-number RANS model used must be able to account for the effect of wall roughness, and some low Reynolds number RANS models are available that can achieve this, e.g. the Spalart-Allmaras model and the k-ω model.

A special situation where wall functions are not well suited and hybrid modelling with RANS in the attached near-wall layers plays an important role are flows around bluff bodies in which large-scale structures are important and shedding may occur. This type of flows is very relevant to hydraulics and may occur e.g. past bridge piers, abutments, groynes, etc. and also in shallow flow with complex bathymetry and past islands. Resolution of the large-scale structures in the detached flow is important for simulating the entire flow properly. The structures do not depend strongly on the Reynolds number. Moreover, the development of these energetic eddies is controlled by instabilities developing outside the attached boundary layers, and resolving accurately the turbulent eddies in the near-wall region having much smaller scales, which of course depend on the Reynolds number, is not so critical. A main goal of such a hybrid-model simulation will be to accurately capture the unsteady dynamics of the large-scale energetic eddies. The number of grid points in the directions parallel to the solid surface is close to independent of the Reynolds number, which greatly reduces the computational cost compared to a wall-resolved LES. These facts led to the introduction of the most popular type of hybrid RANS-LES method, namely the Detached Eddy Simulation (DES) which was originally developed specifically for separated flows. This method and variants derived from it will be described in detail in section 7.4.

7.1.2 Similarity between LES and URANS equations and difference between the approaches

The governing equations in RANS and URANS are the time-filtered Navier-Stokes-equations and in LES the space-filtered Navier-Stokes equations. The equations are formally the same; they are given for LES as Equations (2.9) and (2.10). For RANS/URANS calculations, the resolved quantities would simply carry in addition the superscript RANS attached to the overbar (see Eqn. 2.4)[1] and τ_{ij}^{LES} would be replaced by τ_{ij}^{RANS}. What is different between the unsteady RANS and LES methods is the way in which the stresses τ_{ij}^{RANS} and τ_{ij}^{LES} expressing the effect of the unresolved turbulent fluctuations are calculated. Since all the models we consider are based on the eddy-viscosity concept, this boils down to how the eddy viscosity appearing in the equations is determined. It is the model for the eddy viscosity which determines the solution of the formally identical governing equations. The eddy viscosity is proportional to a typical velocity scale and a typical length-scale of the unresolved turbulent fluctuations. In RANS models, both scales are determined by the mean flow field (e.g. velocity gradients, flow widths) and/or statistical turbulence quantities (e.g. turbulent kinetic energy, turbulent dissipation rate, etc.), while in the case of LES the velocity scale is generally related to gradients of the resolved velocity or to SGS turbulence variables (e.g. SGS kinetic energy), but the length scale is taken as proportional to the local grid size.

Generally, RANS models produce a large eddy viscosity except very close to the wall, which for steady boundary conditions in most cases results in a steady solution. Exceptions are situations with strong instabilities such as the large-scale vortex shedding developing behind bluff bodies. The instabilities responsible for the shedding are not damped out by the eddy viscosity when a suitable turbulence model is used. The shedding motion has a time scale considerably larger than the turbulent time scale and is not considered as part of the turbulence (triple decomposition). This is one case where in a classical sense one would speak of an Unsteady RANS (URANS) calculation and typical examples are flows past bluff bodies as they occur often in hydraulics (e.g. bridge piers). Another classical URANS situation is when the boundary conditions are unsteady, the time scale of the unsteadiness being considerably larger than the turbulent time scale (scale separation). Examples in river and coastal engineering are cases in which large-scale flow unsteadiness is due to external forcing, like in tidal estuary flows or flood-wave events. In both URANS situations, the stresses due to what is considered turbulence are determined by the RANS model. Such calculations should in general be grid and time-step independent. However when the unsteadiness of the boundary conditions is of the time scale of the turbulent fluctuations (e.g. what we can have at an interface where an LES provides the unsteady boundary conditions to the RANS region), there is no clear scale separation. The same is true for RANS models that produce lower eddy viscosity the more instabilities prevail, like in the SAS model described in section 7.4.5, and in this case there is of course a strong time-step dependence determining how much of the fluctuations is resolved and how much is covered by the model. In LES, the eddy viscosity, which is generally related to the mesh size, is much smaller than what is produced by a usual RANS model. Hence, fluctuations are damped much less and in fact most fluctuations contributing to the stresses are resolved up to the dissipative range. When the mesh size is made sufficiently small, a DNS results and all fluctuations are resolved. When on the other hand a fairly coarse grid is used, as in a method sometimes called Very Large Eddy Simulation (VLES), then of course a smaller part of the fluctuations is resolved and

---

1 In the remainder of this chapter, the superscript RANS is omitted so that a quantity with overbar is a time-filtered or space-filtered quantity depending on whether the calculations are carried out in RANS/URANS or LES mode.

more need to be modelled, but doing this with an usual SGS model designed for LES is not advisable. An extreme case is the calculation of flow in large physical domains (large rivers, coastal areas, oceans) with very coarse grids. Here a URANS simulation seems more meaningful than a LES with an SGS eddy viscosity related to the grid size.

7.1.3   Types of hybrid models covered

Various classifications for hybrid models exist in the literature. Models are classified as either zonal or non-zonal depending on whether the regions covered by RANS and LES are predetermined or an outcome of the solution. On the other hand, Fröhlich and von Terzi (2008) – hereafter denoted FT – differentiated in their review paper between unified and segregated hybrid models, the first ones using the same equations for the flow field throughout while in the second ones stand-alone LES and RANS computations are performed in the respective sub-domains. However, none of the classification schemes proposed is unambiguous and allows a clear-cut allocation of the most important models into one or the other category. Hence, here no classification is used but in the following simply the most important and popular types of hybrid models are introduced and described. These are:

–   *Two-layer models* – Here the near-wall region is calculated by RANS and the outside region away from walls by LES, using different eddy viscosity models in the two regions. The interface between the regions located generally parallel to the wall can either be sharp or smeared in the case of blending of the models.
–   *Embedded LES* – Here only a sub-region of the flow is calculated by LES (e.g., see Fig. 7.1) and the rest, when the boundary conditions are steady, by steady RANS. The coupling at the interfaces is here the main problem. The sub-region needs not be treated by a true LES. A hybrid model such as DES or a two-layer model can be used.
–   *Detached Eddy Simulation (DES)* – Here again the near-wall region is calculated by RANS and regions away from the wall and in particular separated flow regions by LES. The same basic eddy-viscosity model is used in both RANS and LES regions - only the length-scale determination is different. In RANS the length scale is either related to the wall distance or is calculated from a length-scale determining model equation, while in the LES region it is taken as the grid size. This model was originally developed for flows with massive separation, but later versions were adapted to make it applicable also to attached flows.
–   *Scale-Adaptive Simulation (SAS) model* – This uses in unsteady 3D calculations a RANS turbulence model that produces in flows with natural fast growing instabilities and unsteadiness, an eddy viscosity so low that larger scales are broken up into smaller ones, reducing further the eddy viscosity until the grid limit is reached. Then, the model switches to an LES mode in which the eddy viscosity is related to the grid size.

A further hybrid model which is not dealt with in detail below is the Partially Filtered Navier-Stokes (PANS) model of Girimaji (2006). This also uses a base RANS model in 3D unsteady calculations, reducing the eddy viscosity such that flow instabilities can grow and unsteady structures prevail. This is achieved by modifying the RANS-

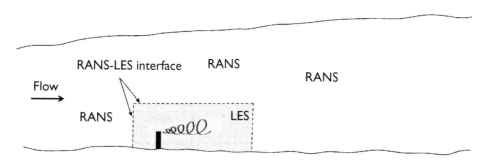

Figure 7.1 Sketch showing a typical example where an embedded LES model is recommended to calculate the flow in a river reach containing a hydraulic structure (wall abutment at the right bank) that induces the formation of a region in which large-scale unsteady eddies are present.

model constants through the prescription of a constant damping ratio throughout the whole computational domain for each characteristic scale of the turbulence closure. The biggest problem is the specification of the damping ratios and it is unclear at the moment whether a general method can be developed that allows the application to complex flows without any prior calibration of the damping ratios.

7.1.4 Numerical requirements

As a simulation performed using a hybrid RANS-LES method contains LES regions, the numerical-scheme requirements, in terms of the properties of the discretization schemes needed to accurately predict the mean flow and turbulence in the LES regions, are identical to those in full LES simulations. Ideally, central difference discretizations, as opposed to low-order upwind discretizations or any extra added numerical dissipation, should be used in these regions. On the other hand, the use of central difference discretizations in the near-wall parts of the RANS regions is not possible due to convergence problems induced by the large stiffness of the discretized system of equations. A good compromise is to use a high-order upwind biased scheme in the RANS regions and to transition smoothly to a central difference scheme in the LES regions. For example, in the case of DES, the expressions of the blending functions involved can be found in Travin et al. (2000). The use of the first order upwind scheme to discretize the convective terms in the near-wall regions is not acceptable. In many cases, the use of a fifth-order upwind biased scheme over the whole domain is acceptable. However, careful validation of the code is needed for each class of applications, at least for simpler but representative flows for which validation data are available.

7.2 TWO-LAYER MODELS

As was mentioned already, in these models a layer near the wall is calculated with a RANS model and the region away from walls by LES. The RANS and LES solutions are either coupled at a sharp interface between the two regions or are blended at a smeared interface. Different models for determining the eddy viscosity are used in the two regions with its value and also the values of other quantities matched at the interface. An overview of the early developments is given in Cabot and Moin (1999) and Piomelli and Balaras (2002). The Detached Eddy Simulation (DES) method introduced in section 7.4 is basically also a two-layer model, but here the same basic eddy-viscosity model is used in the two regions, with only the length scale appearing in the transport equations for the quantities used to estimate the eddy viscosity determined differently.

7.2.1 Models with a sharp interface between RANS and LES regions

Of prime importance in a two-layer model is the location of the interface between the RANS and LES regions. A so-called hard interface results when the location is predetermined. When structured grids are used, the interface is often placed at a certain grid line which has, on average, a certain non-dimensional distance $z^+ = zu_*/\nu$ from the wall. Typical values chosen are of the order of 100. With "soft" interfaces the location is an outcome of the solution, generally controlled by choosing again a certain constant value of z^+. As the friction velocity u_* varies with time, so does the wall distance at the interface, z, in this case. One problem with the relation of the interface location to the bed friction velocity is that the latter becomes very small close to separation and reattachment. Though one can empirically fix this problem by placing a lower bound on the value of u_* used to estimate z^+, one can also look for other approaches (e.g., Breuer et al., 2007, Kniesner et al., 2007) that do not face these limitations. For example, Breuer et al. (2007) proposed to use the modelled turbulent kinetic energy, k, to define the position of the interface (e.g., $z'^+ = z\sqrt{k}/\nu = const.$).

A few examples of two-layer models with sharp interface will now be presented. Davidson and Peng (2003) used the k-ω model in the RANS sub-domain and a k-equation SGS model in the LES sub-domain. The equation for the turbulent kinetic energy k was solved continuously across the interface. The ω-equation was solved only in the RANS sub-domain with a zero normal gradient boundary condition for ω at the interface position. A similar approach was used by Temmerman et al. (2005). The main difference was that a one-equation model for the turbulent kinetic energy was used in the RANS sub-domain. Kniesner et al. (2007) used a k-ε model in the RANS sub-domain and either the Smagorinsky model or a one-equation SGS model in the LES sub-domain. When the classical Smagorinsky model was used, the values of k and ε at the interface were evaluated based on the resolved flow variables in the LES sub-domain.

In such models, unphysical flow structures can be generated around the location of the interface, which negatively affects the prediction of the mean flow and turbulence statistics. For example, the unphysical flow structures generally induce a mismatch of the slopes of the logarithmic velocity profile in calculations of turbulent channel flow (see also Hamba, 2003). In a simulation of the flow over periodic hills, Temmerman et al. (2005) observed the formation of kinks in the mean velocity profiles around the interface location. Still, the mean flow predictions were in much better agreement with well-resolved LES than those given by RANS. An overview of the performance of two-layer models for calculating the periodic hill flow test case is given in Fröhlich and von Terzi (2008).

An important sub-category of two-layer models are models based on the simplified Turbulent Boundary Layer (TBL) equations (Balaras et al., 1996, Cabot and Moin, 1999,

Piomelli and Balaras, 2002) within the near-wall layer. These equations, together with a RANS model (usually an algebraic eddy-viscosity model), are employed to compute the instantaneous wall shear stress, which is then used as approximate boundary condition for LES. Two separate grids are used in two-layer models based on the TBL equations (Fig. 7.2). The coarser LES grid covers the full domain starting at the walls. The TBL equations are solved on an embedded near-wall mesh that is finer in the wall-normal direction only. The TBL equations are forced at the boundary between the RANS (wall layer) and LES (outer layer) regions by the instantaneous tangential velocities from LES. No-slip conditions for the velocity are applied at the wall. Thus, the flow in the wall layer is unsteady. However, in some simplified versions of this type of models, the unsteady term in the TBL equations is neglected (e.g., see discussion of the Wang and Moin, 2002, model below). These methods were successful in calculating channel flow at high Reynolds numbers and backward-facing-step flow (Cabot, 1996, Diurno et al., 2001).

The model of Wang and Moin (2002) based on solving the TBL equations in the near-wall region will now be described. In this model the coefficients in the RANS model are adjusted dynamically so that RANS eddy viscosity and SGS viscosity predicted by LES match at the RANS-LES interface (see discussion of Eqn. 7.4). The values of the eddy viscosity in the near-wall layer are reduced compared to the typical RANS values in order to account only for the unresolved part of the total turbulent stress in the wall layer. This is achieved by using a dynamically adjusted mixing-length eddy viscosity in the TBL equations.

Assuming that in a local system of coordinates $i = 3$ represents the wall normal direction, the TBL momentum and continuity equations read:

$$\frac{\partial}{\partial x_3}\left[(v+v_t)\frac{\partial \overline{u}_i}{\partial x_3}\right] = F_i \quad i = 1 \quad \text{and} \quad i = 2, \quad \text{where} \quad F_i = \frac{1}{\rho}\frac{\partial \overline{p}}{\partial x_i} + \frac{\partial \overline{u}_i}{\partial t} + \frac{\partial \overline{u}_i \overline{u}_j}{\partial x_j} \tag{7.1}$$

$$\frac{\partial \overline{u}_i}{\partial x_i} = 0 \tag{7.2}$$

The turbulent viscosity is obtained using a mixing-length model with wall damping:

$$\frac{v_t}{v} = \kappa z^+ (1 - e^{-z^+/A^+})^2 \tag{7.3}$$

where z^+ is the distance to the wall in wall units, κ is the main model coefficient (in the original RANS mixing-length model $\kappa = 0.4$ is von Karman's constant) and $A^+ = 19$.

The pressure present in F_i is assumed not to vary significantly in the wall-normal direction x_3 and is taken equal to the value from the outer-flow LES solution (first point off the wall in the LES grid). Equations (7.1) are required to satisfy the no-slip conditions at the wall and to match the outer layer solutions at the first off-wall LES velocity grid points on the coarser LES grid.

In general, the full TBL Equations (7.1) and (7.2) are solved numerically to obtain $\overline{u}_1$ and $\overline{u}_2$ within the wall layer and the wall-shear-stress components τ_{w_1} and τ_{w_2} (Fig. 7.2). The TBL equations are integrated in time along with the outer-flow LES equations.

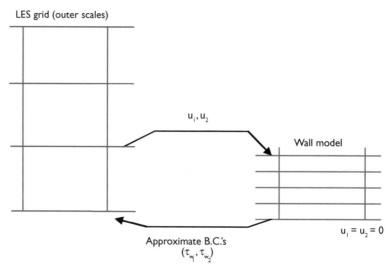

Figure 7.2 Sketch showing the inner and outer grids for the two-layer model of Wang and Moin (2002).

The wall-normal velocity component $\bar{u}_2$ in the near-wall-layer region is determined from the continuity Equation (7.2). No Poisson equation is required to be solved in the inner layer, since the pressure is assumed constant in the wall-normal direction.

Because the unsteady velocities are matched at the boundary between the LES and RANS regions, the TBL region contains resolved unsteady motions which generate stresses. Hence the RANS model should account only for the unresolved part of the total turbulent stresses. Thus, the eddy viscosity predicted by the model used in the near-wall layer should be lower than the values predicted in full RANS simulations. To achieve that, Wang and Moin (2002) allowed the model coefficient κ to vary such that the mixing-length (RANS) eddy viscosity and the SGS viscosity are equal at the interface. The model coefficient is calculated using Equation (7.3) at the interface as

$$\kappa = <V_{SGS_s}>/<vz^{+}(1 - e^{-z^{+}/A^{+}})^{2}> \tag{7.4}$$

where the brackets <> denote averaging in time (of the order of 100 time steps back in time) and the subscript 's' the value at the interface. The averaging is needed to reduce the point to point oscillations in the model coefficient. For attached turbulent boundary layers, the value predicted by the two-layer model was around 0.1, which is significantly lower than the usual value of von Karman's constant ($\cong 0.4$).

In the TBL region, the grid spacing in the wall-normal direction is recommended to be smaller than 3 wall units, with the first point off the wall situated at approximately one wall unit. The first point off the wall in the LES grid is generally situated at 30 to 200 wall units. The computational cost to solve the TBL equations is insignificant compared to the outer layer LES as there is no need to solve the momentum equation in the wall normal direction and the Poisson equation for the pressure. Also, the TBL equations are much simplified in the locally orthogonal wall-layer coordinates as no cross-derivative terms are present.

There are several simpler variants of the model described above. They are obtained by retaining only the pressure term in F_i on the right hand side of Equation (7.1), or simply by setting $F_i = 0$, in which case the new model was called the equilibrium stress balance model. In the latter case, one can show that the algebraic model implies the logarithmic law of the wall for the instantaneous velocities for $z^+ \gg 1$, and the linear velocity distribution for $z^+ \ll 1$. Moreover, Equation (7.1) can be integrated numerically directly between the wall and the first point off the wall in the LES grid, without the need to build a wall-layer grid, which greatly simplifies the implementation of the wall-layer model in complex geometries. However, retaining the pressure term in F_i was found to give clearly superior results.

The full model of Wang and Moin (2002) and in most cases also the simplified version keeping only the pressure term in F_i were shown to predict low-order statistics (e.g., mean velocities, Reynolds stresses) in very good agreement with those from well-resolved LES, at a much smaller computational cost (more than one order of magnitude lower compared to that of a well-resolved calculation) for several complex turbulent flows. The model was used to compute several types of complex flows including backward facing step, airfoil at angle of attack, cylinders at high Reynolds numbers (Catalano et al., 2003) and, recently, flow in a meandering channel with natural bathymetry (Kang and Sotiropoulos, 2011).

7.2.2 Models with a smooth transition between RANS and LES regions

In this type of models a blending function f is used to determine the value of a certain turbulence variable ϕ (eddy viscosity, turbulent kinetic energy, etc.) from the values given by the RANS (ϕ^{RANS}) and LES (ϕ^{LES}) components of the model, i.e. the value of ϕ is calculated as the weighted sum of these two components:

$$\phi = f\phi^{RANS} + (1-f)\phi^{LES} \tag{7.5}$$

In general, f is a continuous monotonous function ($0 < f < 1$) determined empirically, which is not necessarily the same for all the variables for which blending is applied. The blending function mainly depends on the wall distance, so that the influence of the RANS model is restricted to near-wall layers.

Fan et al. (2004) proposed a hybrid model in which the k-ω model was blended with a SGS model. In the LES region, the unresolved kinetic energy was determined from a transport equation SGS model (see section 3.3.3). The model blended the values of the eddy viscosity and of the dissipation rate in the transport equation for the turbulent (subgrid) kinetic energy. The shape of the blending function f was similar to that used in the standard SST model of Menter (1994) and was a function of the wall distance. Xiao et al. (2004) proposed several variants of the same model, including using different blending functions and a different RANS model. The model was used with moderate success to predict flow over a ramped cavity. A drawback of this approach is that unphysical structures may be generated in simulations using blending models. A typical example is that of turbulent channel flow (e.g., see Baggett, 1998) where such simulations significantly overpredict the size of the streamwise velocity streaks. The presence of these unphysical structures, sometimes called super-streaks, in simulations using blended models results in the formation of a spurious buffer layer and errors in the prediction of the mean velocity profile.

7.3 EMBEDDED LES

This type of modelling is suitable when major parts of the entire computational domain can be calculated with sufficient accuracy by an economical RANS model (usually steady) and only a sub-domain, where the flow is too complex to be captured by RANS, needs to be simulated by LES. An example of such a situation is shown in Figure 7.1, namely the flow in a river with a groyne structure present at one of the river banks. Here the separated flow past the groyne is controlled by the dynamics of large-scale unsteady structures or instabilities and requires for proper calculation an eddy-resolving method, while the rest of the river can be calculated by a RANS method. The LES need not be a pure LES but can itself be a hybrid method (e.g. DES) in order to overcome the resolution problem at high Reynolds numbers near the wall boundaries of the LES sub-domain (groyne walls and river banks in Fig. 7.1). As can be seen from Figure 7.1, in an embedded LES the LES region is bordered by RANS regions. The main problem is the coupling of the solutions obtained in the different regions. In principle, the coupling should always go both ways across the interface, but the simpler one-way coupling may often be sufficient. As discussed by Fröhlich and von Terzi (2008), depending on the direction of the mean flow with respect to the interfaces between the RANS and LES sub-domains, several relevant situations can occur. The boundary surfaces that are generally the most critical for the accuracy of the embedded LES simulation are the inflow to the LES sub-domain (transition from RANS to LES) and the outflow of the LES sub-domain (transition from LES to RANS).

7.3.1 Inflow to LES sub-domain

At the upstream boundary of the embedded LES domain, the steady or unsteady RANS solution provides the distribution of the mean velocity and of some of the turbulence variables (e.g., the turbulent kinetic energy). The RANS solution does not provide any turbulent fluctuations. However, for most flows the quality of the LES solution depends strongly on whether or not velocity fluctuations with properties that are as close as possible to those of the real turbulent flow are present at the inflow boundary of the LES sub-domain. The task is then to generate at this boundary unsteady velocity fields with realistic turbulent fluctuations whose statistics are compatible with the RANS solution. The latter requirements makes the generation of inflow conditions different from that for a pure LES described in chapter 6, but similar methods can be employed.

Two main types of methods can be used to provide unsteady fluctuations. The first type provides physical flow fields containing realistic flow structures with the aid of a precursor calculation, while the second type generates synthetic turbulent fluctuations having a certain spectrum. The information from RANS on the distribution of the turbulence variables at the inflow section can be used to rescale the unsteady fluctuations such that their statistics (e.g., turbulent kinetic energy) are consistent with those predicted by RANS.

In the precursor calculations of the first type of methods, eddy-resolving simulations are performed to obtain the turbulent fluctuations which are then stored in a database. In most practical applications in hydraulics, the flow upstream of the LES domain is that in an open channel, often with irregular cross section. Therefore, a precursor calculation with streamwise periodic boundary conditions of developed flow in a straight channel having the same cross section is suitable in most cases. In general, the Reynolds number is so high that a well-resolved precursor LES is not possible. Hence the calculations have to be carried out either with wall functions or with a hybrid model (two-layer approach). Alternatively, the velocity fluctuations can be obtained from a well-resolved LES at a relatively low Reynolds number. Then, the fluctuations are rescaled and added to the mean-flow field provided by RANS. In order that the mean flow and, if desired, also the turbulence statistics of the precursor LES match closely those of the RANS calculations, additional forcing terms and a controller have to be used in the precursor LES. The details of such methods are discussed in Pierce (2001) and Garcia-Villalba and Fröhlich (2006). A special adaptation of the methods to the hybrid approach, introducing an overlap zone at the interface, is described in Keating et al. (2004).

Another option is to recycle flow structures predicted within the LES sub-domain some distance from the inflow section (e.g., see Lund et al., 1998, Schluter et al., 2004, 2005). This is an option only if the shape of the LES sub-domain does not change too much over the upstream part of the LES sub-domain. Some additional rescaling may be needed. As the mean-velocity field at the inlet section is known from RANS, only the velocity fluctuations need to be recycled and rescaled.

The computational cost of the precursor simulation can be substantial. If precursor eddy-resolving simulations cannot be performed (e.g., the flow at the inflow section of the LES sub-domain is very different from a fully developed channel flow) or are too expensive, one can use methods that generate synthetic turbulent fluctuations based on POD modes, Fourier modes, stochastic forcing, digital filtering (see brief description in section 6.3.2), etc. These methods (e.g., see Batten et al., 2004, Druault et al., 2004, Klein et al., 2003) are computationally much less expensive and more general in terms of the characteristics of the turbulent fluctuations that are generated. A further possibility to supplement the mean velocity discussed in section 6.3.2 is by introducing physical flow structures in form of random vortices at the interface between the RANS and LES regions. Such methods have been developed and tested by Mathey et al (2003), Lefevre (2004) and Mathey and Cokljat (2005). As reported in von Terzi and Fröhlich (2008) for the case of flow over periodic hills, these methods produced significantly more accurate results compared to methods that supplement the mean velocity by random noise.

An important observation is that the turbulent fluctuations obtained using these methods have a spectrum that approximates the one observed in real turbulent flows. This is why the fluctuations generated using these methods should develop relatively rapidly into 'realistic' turbulence. The more realistic the fluctuations generated by the synthetic turbulence method are, the better is the chance that these fluctuations generate physical turbulent eddies. The use of random noise superimposed on the mean-flow field predicted by RANS generally results in a rapid damping of the fluctuations. Thus, imposing fluctuations with characteristics that are as close as possible to those present in the corresponding real turbulent flow is of utmost importance.

7.3.2 Outflow from LES sub-domain

At the downstream boundary of the embedded LES region the formulation of consistent boundary conditions can be difficult if two-way coupling is used. This is because the interface should allow for the mean-flow information to be propagated upstream, from the RANS region toward the LES region. In hydraulics the flow is incompressible so that one can completely decouple the pressure between the LES and RANS regions and only couple the velocity fields. Thus, the discussion below considers two-way and one-way coupling procedures only for the velocity fields, but the emphasis is on one-way coupling.

In the most general case, the outflow boundary conditions should be formulated to allow the RANS and LES sub-domains to exchange mean flow information, to allow the smaller-scale LES fluctuations to leave the LES sub-domain without reflections and to allow estimating the turbulence variables for which transport equations are solved in the RANS sub-domain.

Whether or not a two-way coupling procedure is needed depends on the specific flow geometry, flow physics and the position of the downstream boundary of the LES region. For example, if in a river reach the boundary between the LES sub-domain and the RANS sub-domain is situated sufficiently far from the region where large-scale strongly energetic eddies originate (e.g., around the hydraulic structure present inside the LES sub-domain), then two-way coupling is not really necessary and using one-way coupling is an acceptable approximation.

In the case of one-way coupling, the standard methods used to specify the boundary conditions at the outflow section in full-domain LES simulations as described in chapter 6 can be used for the LES sub-domain. The main requirement is that the fluctuations leave the LES domain without reflections. The most robust and popular method is the convective outflow boundary condition given in Equation (6.2). Then, to achieve the one-way coupling one has to calculate from the LES solution the statistical mean-flow and turbulence quantities that need to be specified at the upstream boundary of the RANS region situated downstream of the LES sub-domain.

When an unsteady calculation is performed in the RANS region, i.e. a URANS simulation, small-scale fluctuations fed into this region from the upstream LES sub-domain may have to be damped out. This can be achieved by placing a buffer (overlap) region close to the outflow from the LES domain within which the smaller scale LES eddies are damped out (e.g., see Freund, 1997, Israeli and Orszag, 1981).

Von Terzi and Fröhlich (2007) proposed a general method for coupling at the downstream interface of an LES domain in which the Reynolds-averaged velocity field of the LES domain was imposed as a Dirichlet condition at the inflow of the RANS domain, while a discrete analogue of a convective outflow boundary condition was prescribed for the velocity perturbation exiting the LES domain.

7.3.3 Lateral coupling of LES and RANS sub-domains

The case of lateral (tangential) coupling at interfaces more or less aligned with the mean flow direction is either analogous to near-wall modelling of LES using a two-layer approach like the one discussed in section 7.2.1., if the interface is close to solid surfaces, or to coupling in the outer-flow region. In the former case, fluctuations have to be provided at the tangential boundary of the LES domain. In the latter case, methods similar to those used at the downstream boundary of the embedded LES sub-domain can be generally used.

7.4 DETACHED EDDY SIMULATION (DES) MODELS

7.4.1 Overview of DES model

Detached Eddy Simulation (DES) was originally intended as a method for simulation of high-Reynolds-number massively separated flows in which the instabilities in the detached shear layers are very strong (e.g., the formation of the large vortices is due to the rapid growth of the Kelvin Helmholtz instability) and their development is relatively independent of the turbulence inherited from the attached boundary layers (Spalart, 2009). For these flows, the solution has little sensitivity to the boundary-layer turbulence. The argument made by Spalart et al. (1997), who proposed DES, was that the momentum transfer far from the walls is dominated by large 'detached' unsteady eddies which are typically geometry dependent and could be resolved by LES without the high grid resolution needed in LES that resolves well the near-wall turbulent structures, including the wall streaks in the attached near-wall flows. Consequently, away from the walls the largest part of turbulent stresses will be calculated directly and will be anisotropic. Especially for massively separated flows, much of the burden of predicting the Reynolds stresses is shifted from the turbulence model to the explicit averaging of a time-dependent solution that resolves the dynamically important eddies. It should also be pointed out that various versions of DES are now successfully applied to other categories of turbulent flows (Spalart, 2009).

In the DES method, the base model is a RANS model and the most important modification needed to produce the switch of this model to LES mode is to relate the turbulent length scale appearing in the transport equations solved for calculating the turbulence variables in the RANS model to the grid size. More specifically, DES uses a RANS closure in the attached near-wall layers and a SGS model in which the eddy viscosity is proportional to the square of the local grid size away from the walls. There is a single solution field, and the transition between the RANS and LES regions, all coupled by the Navier-Stokes equations, is seamless in an application sense, i.e., without artificial transitions between the different solution domains. Compared to URANS, DES is more expensive for a fully three-dimensional flow, as the time steps in the simulation need to be smaller and the grid in some regions of the flow has to be refined to capture the dynamically important eddies in that region. However, the overall increase is generally less than one order of magnitude and the accuracy of the DES predictions improves in most cases significantly compared to URANS (e.g., see Constantinescu et al., 2003). Moreover in DES, as in classical LES, increased spatial and temporal resolution yields additional information on the physics and the fine-grid limit of DES is DNS.

The original version of DES (Spalart et al., 1997) is based on a modification of the turbulent length scale in the destruction term of the model transport equation for eddy viscosity in the Spalart-Allmaras (1994) RANS model. Versions using other RANS models were proposed, in particular the k-ω SST model proposed by Menter (1994). The modifications needed to obtain DES formulations based on these two RANS models are described next. Regardless of the DES formulation used, the first grid point off the wall should be situated within the viscous sub-layer ($z^+ \cong 1$ is recommended).

7.4.2   DES based on the Spalart-Allmaras (SA) model

RANS model

The SA model is based on the following transport equation for the modified eddy viscosity $\tilde{V}$ (its relation to v_t is given in Eqn. 7.7 below). The transport equation for $\tilde{V}$ is:

$$\frac{\partial \tilde{V}}{\partial t} + \bar{u}_j \frac{\partial \tilde{V}}{\partial x_j} = C_{b1}\tilde{S}\tilde{V} + \frac{1}{\sigma}\left[\nabla \cdot ((v + \tilde{V})\nabla \tilde{V}) + C_{b2}(\nabla \tilde{V})^2\right] - C_{w1}f_w\left[\frac{\tilde{V}}{d}\right]^2 \tag{7.6}$$

and the eddy viscosity v_t is obtained from

$$v_t = \tilde{V}f_{v1} \tag{7.7}$$

The destruction term in the $\tilde{V}$-equation contains the turbulent length scale, d, which for a smooth wall is equal to the distance to the nearest wall, d_{min}.

The other variables ($\bar{S}$ is the magnitude of the resolved vorticity vector; alternatively $\bar{S}$ can be defined as the magnitude of the resolved rate of strain tensor without a change in the model constants) and parameters in Equations (7.6) and (7.7) are given by:

$$\tilde{S} \equiv \bar{S} + (\tilde{V}/\kappa^2 d^2)f_{v2}; \quad f_{v2} = 1 - (\tilde{V}/v)/(1 + \chi f_{v1}); \quad f_{v1} = \chi^3/(\chi^3 + C_{v1}^3)$$

$$\chi = \tilde{V}/v + 0.5\frac{k_s}{d}; \quad f_w = g\left[\frac{1 + C_{w3}^6}{g^6 + C_{w3}^6}\right]^{\frac{1}{6}}; \quad g = r + C_{w2}(r^6 - r); \quad r \equiv \frac{\tilde{V}}{\tilde{S}\kappa^2 d^2} \tag{7.8}$$

where k_s is the equivalent roughness height (which can include the form roughness due to the presence of ripples or small dunes that are not resolved by the grid in simulations of flow in natural channels). The model constants are: $C_{b1} = 0.135$, $C_{b2} = 0.622$, $\sigma = 0.67$, $\kappa = 0.41$, $C_{v1} = 7.1$, $C_{w2} = 0.3$, $C_{w3} = 2.0$, $C_{w1} = C_{b1}/\kappa^2 + (1 + C_{b2})/\sigma$.

To account for roughness effects, the distance to the (rough) wall is redefined (see also Spalart, 2000) as:

$$d = d_{min} + 0.03k_s \tag{7.9}$$

For smooth walls, k_s and $\tilde{v}$ at the wall are taken as zero. For rough walls, the value of $\tilde{v}$ at the wall is estimated by applying the boundary condition $\partial \tilde{v}/\partial n = \tilde{v}/d$ where n is the direction normal to the wall (Spalart, 2000). This makes the modified viscosity and the eddy viscosity to be formally non-zero at the rough walls.

Spalart and Allmaras (1994) present additional modifications of the transport equation needed for calculations in which the flow may be non-turbulent is some regions (e.g., in cases when the model is used to calculate transition).

Model in LES mode

The SA version of DES is obtained by replacing the RANS length scale $d_{RANS} = d$ in the destruction term and in the expressions of the model parameters of the SA model by a new length scale

$$\tilde{d} = \min(d_{RANS}, d_{LES}) \tag{7.10}$$

where the LES length scale is defined as

$$d_{LES} = C_{DES}\Delta \tag{7.11}$$

In Equation (7.11), Δ is the local maximum grid spacing in the three directions ($\Delta = max(\Delta_x, \Delta_y, \Delta_z)$). As illustrated in Figure 7.3, the consequence of Equation (7.10) is that in regions near the wall where the wall distance d is smaller than $C_{DES}\Delta$, the SA model runs in RANS mode while further away from the wall, where $C_{DES}\Delta > d$,

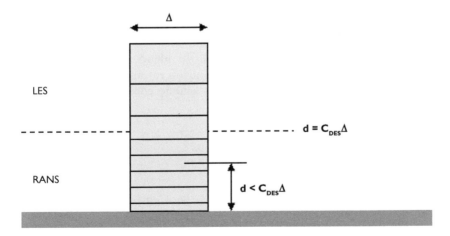

Figure 7.3 Sketch showing the RANS and LES regions in a DES simulation. The boundary corresponds to the location where the definition of the turbulence length scale switches from being proportional to the distance to the closest wall to being proportional to the local grid size, Δ. (adapted from von Terzi and Froehlich, 2008).

the SA model acts as SGS model in a LES. In standard applications of DES, the wall-parallel grid spacing is at least of the order of the boundary layer thickness so that the SA RANS model is effective throughout the boundary layer, i.e., $\tilde{d} = d$. Consequently, prediction of boundary layer separation is determined in the RANS mode of DES.

Far from the walls, $C_{DES}\Delta < d$ (Fig. 7.3), and when turbulence is near equilibrium so that the dominant (production and destruction) terms in the model Equation (7.6) are in balance, there follows $\tilde{\nu} \sim \tilde{S}\Delta^2$ near equilibrium. This means that the eddy viscosity assumes a Smagorinsky-like expression in the 'LES region' once the turbulent length scale becomes proportional to the local grid spacing. The length scale redefinition away from the walls increases the destruction term in the SA model, draws down the eddy viscosity, and allows instabilities to develop. Roughly, it makes the pseudo-Kolmogorov length scale, based on the eddy viscosity, proportional to the local grid spacing. The model constant C_{DES} is of the order of one, and it should be set such that the spectrum at high resolved frequencies does not exhibit oscillations, and the decrease in the spectrum at high frequencies is not too steep, meaning that the turbulent eddies are not under-resolved at these frequencies. The value $C_{DES} = 0.65$ was obtained based on calibration in homogenous turbulence (Shur et al., 1999). Sometimes, smaller values of C_{DES} are used to compensate for the larger numerical diffusivity in the numerical solver used to perform the DES simulations.

7.4.3   DES based on the SST model

RANS model

The k-ω Shear-Stress-Transport (SST) model was proposed by Menter (1994). The SST model employs the classical k-ω model near solid walls and the high-Reynolds-number version of the k-ε model away from walls and in free-shear layers. Major features of the SST model are the blending of the two models by zonal weighting of the model coefficients and limiting the eddy viscosity in adverse-pressure gradient boundary layers. To blend the k-ω and k-ε models, the latter is rewritten using a transformation of variables into a k-ω like form. The transport equations for the turbulent kinetic energy, k, and the turbulent vorticity, ω, in the SST model are:

$$\frac{\partial k}{\partial t} + \bar{u}_j \frac{\partial k}{\partial \bar{x}_j} - \frac{\partial}{\partial x_j}\left((\nu + \nu_t \sigma_k)\frac{\partial k}{\partial x_j}\right) = G - \beta^* \omega k \tag{7.12}$$

$$\frac{\partial \omega}{\partial t} + \bar{u}_j \frac{\partial \omega}{\partial x_j} - \frac{\partial}{\partial x_j}\left((\nu + \nu_t \sigma_\omega)\frac{\partial \omega}{\partial x_j}\right) = \frac{\gamma}{\nu_t}G - \beta \omega^2 + 2(1 - F_1)\frac{\sigma_{\omega 2}}{\omega}\frac{\partial k}{\partial x_j}\frac{\partial \omega}{\partial x_j} \tag{7.13}$$

The production term G is defined as:

$$G \equiv \frac{1}{2}\nu_t\left(\frac{\partial \bar{u}_i}{\partial x_j} + \frac{\partial \bar{u}_j}{\partial x_i}\right)^2 \quad \text{(summation over } i \text{ and } j) \tag{7.14}$$

The blending function F_1 is designed to have a value of one in the near wall region (k-ω mode) and a value of zero in the regions further away from the walls (k-ε mode). Its expression is:

$$F_1 = \tanh\left\{\min\left[\max\left(\frac{\sqrt{k}}{0.09\omega d}, \frac{500\nu}{d^2\omega}\right); \frac{4\sigma_{\omega 2}k}{CD_{k\omega}d^2}\right]^4\right\}$$
(7.15)

where $CD_{k\omega} = \max\left[\dfrac{2\sigma_{\omega 2}}{\omega}\dfrac{\partial k}{\partial x_j}\dfrac{\partial \omega}{\partial x_j}; 10^{-20}\right]$. The eddy viscosity is defined as:

$$\nu_t = \frac{k/\omega}{\max\left[1; \overline{\Omega}F_2/(a_1\omega)\right]}$$
(7.16)

where $a_1 = 0.31$ and $\overline{\Omega} = \sqrt{\overline{\Omega}_{ij}\overline{\Omega}_{ij}}$ and $\overline{\Omega}_{ij}$ is the mean rate of rotation tensor. For the particular case of a boundary layer, the quantity $\overline{\Omega}$ is basically the mean velocity gradient in the wall-normal direction. The eddy viscosity formulation given by Equation (7.16) guarantees that in adverse-pressure gradient boundary layers, where production of k can be significantly larger than dissipation, the shear stress is calculated from $\tau = \rho a_1 k$ and thereby limited, while in other flows (including free shear layers), the original eddy-viscosity relation $\nu_t = k/\omega$ applies. The blending function F_2 serves to restrict the modification to wall-bounded flows. Its expression depends therefore on the modified wall distance, d and reads:

$$F_2 = \tanh\left\{\left(\max\left[2\frac{\sqrt{k}}{0.09\omega d}; \frac{500\nu}{d^2\omega}\right]\right)^2\right\}$$
(7.17)

The model constants are $\beta^* = 0.09$ and $\kappa = 0.41$. The coefficients $\phi = \{\sigma_k, \sigma_\omega, \beta, \gamma\}$ of the SST model are defined by blending $(\phi = F_1\phi_1 + (1-F_1)\phi_2)$ the corresponding coefficients of the original k-ε and k-ω models (see Table 7.1)

At the walls, k is set equal to zero. For smooth walls $\omega = 800\nu/(z_1^2)$, where z_1 is the normal distance to the wall of the first grid point off the wall. For rough surfaces $\omega = 2500\nu/(k_s)^2$ for $k_s^+ < 25$ and $\omega = 100u_*/k_s$ for $25 < k_s^+$, where $k_s^+ = u_* k_s/\nu$.

Table 7.1 The blending function coefficients of the SST model.

Inner model coefficients	Outer model coefficient
$\beta_1 = 0.075$	$\beta_2 = 0.0828$
$\sigma_{k1} = 0.85$	$\sigma_{k2} = 1.00$
$\sigma_{\omega 1} = 0.5$	$\sigma_{\omega 2} = 0.856$
$\gamma_1 = \beta_1/\beta^* - \sigma_{\omega 1}\kappa^2/\sqrt{\beta^*} = 0.553$	$\gamma_2 = \beta_2/\beta^* - \sigma_{\omega 2}\kappa^2/\sqrt{\beta^*} = 0.440$

Model in LES mode

The length scale in the k-ω SST model is implicit. To be able to propose a DES version, Strelets (2001) explicitly redefined the turbulent length scale in the RANS model as:

$$d_{RANS} = k^{1/2}/(\beta^* \omega) \tag{7.18}$$

Then, in analogy to Equation (7.10), the length scale in the DES formulation of the SST model is redefined as:

$$d_{DES} = min(d_{RANS}, d_{LES}) \tag{7.19}$$

where d_{LES} is given by Equation (7.11). The only term of the SST model that has to be modified to obtain a DES formulation is the dissipation term in the transport equation for k which now reads $D_{DES}^k = \rho k^{3/2}/d_{DES}$. Separate calibrations of the model constant, C_{DES}, can be performed for the k-ε and k-ω branches of the SST model. Then, these values are blended using the function F_1 (Eqn. 7.15).

$$C_{DES} = (1 - F_1)C_{DES}^{k-\varepsilon} + F_1 C_{DES}^{k-\omega} \tag{7.20}$$

From the standpoint of DES, only the k-ε branch is important, since this branch is active in most of the regions where the DES runs in LES mode. Strelets (2001) proposed using $C_{DES}^{k-\varepsilon} = 0.61$ and $C_{DES}^{k-\omega} = 0.78$.

7.4.4   Improved versions of DES

As already mentioned, the original version of DES was developed for calculating flows with massive separation. To establish DES as a general method for calculating high-Reynolds-number complex flows, DES has to be able to predict with reasonable accuracy also attached flows. Application of the standard DES model as kind of a wall model to boundary layer-type flows with the RANS/LES interface situated inside the boundary layer leads to a grey (or transition) area around the interface with mismatch of the log laws on either side, generally insufficient fluctuations in the LES region and unphysical structures forming a spurious buffer layer. Problems are encountered especially when ambiguous grids with wall-parallel grid spacing of the order of the boundary layer thickness are used, in which case the phenomena of Modelled Stress Depletion (MSD) and Grid-Induced Separation (GIS) can occur. These problems are discussed further in the next sections together with remedies proposed to overcome them.

DES with spectral enrichment of the near wall flow

In channel flows, the standard SA and SST versions of DES predict a too high velocity gradient in the region where the model transitions from RANS mode to LES mode. This is due to the formation of unphysical, large, nearly one-dimensional streaks of high and low streamwise velocity (super-streaks). This region resembles a spurious buffer layer which shows up as a shift in the logarithmic law compared to experiment and well-resolved LES.

The inaccuracy of the DES solution is due to the fact that for attached flows there is no mechanism to generate velocity fluctuations at the rate required for the transition between the quasi-steady-state RANS region and the unsteady LES region and to proceed in a way consistent with the flow physics (e.g., generating velocity fluctuations inside the buffer layer at the right rate). Thus, one possible way to alleviate this problem is to increase the resolved turbulent shear stress in the transition region.

Piomelli et al. (2003) was able to suppress the formation of the super-streaks and to obtain the correct mean streamwise velocity profile in a channel flow DES simulation by the addition of a stochastic forcing term, f_i, to the right hand side of the momentum equations. The term f_i was isotropic, was obtained from a normally distributed random series with a given magnitude, and was active in a region surrounding the RANS/LES interface. The forcing was projected into a divergence-free field before being added to the right-hand side of the momentum equations. A drawback of the procedure is that calibration of the magnitude of the forcing term was needed to remove the shift in the logarithmic law. The effect of the local forcing was to reduce the thickness of the transition region by increasing the transfer of energy form the modelled to the resolved eddies. The forcing acts as a backscatter model that adds energy mainly to the smaller resolved scales present close to the RANS/LES interface. A similar approach was also proposed by Davidson and Dahlstrom (2005). In a following paper, Keating and Piomelli (2006) proposed a procedure to calculate the amplitude of the stochastic forcing dynamically. The authors discuss the details of the implementation of this model for periodic and spatially developing turbulent channel flow, including the effect of the use of synthetic turbulence at the inflow section in the latter case. The method was shown to successfully predict the profiles of the mean velocity and velocity r.m.s. fluctuations for these two flows.

Delayed DES (DDES)

In cases when the attached boundary layer is thin and the wall-parallel grid size is generally larger than the boundary layer thickness, the DES switches from the RANS mode to the LES mode outside the boundary layer, as intended in the original DES proposal. However, if the grid spacing parallel to the wall is smaller than the boundary layer thickness, e.g. in the case of thick boundary layers and shallow separated regions, the original DES formulation tends to lead to the problems indicated already above. The grid spacing in such a case is fine enough to invoke the LES mode of DES (see Eqn. 7.10) over the outer part of the attached boundary layer, lowering the eddy viscosity below levels that would be predicted if the RANS mode will be active in this region. However, the grid is on the other hand not fine enough to produce realistic "LES content", and hence the turbulent stresses caused by the resolved velocity fluctuations are smaller than the Reynolds stresses that would be predicted in the same region if the model would remain in RANS mode. This is the phenomenon called Modelled Stress Depletion. It causes the predicted skin friction to be too small and may cause premature separation.

A solution to this problem within the DES framework was proposed by Menter and Kuntz (2002) who used the blending functions F_1 or F_2 of the SST model to identify the boundary layer, and prevent the early switch to LES. The two blending functions are equal to one inside the boundary layer and fall rapidly to zero away from the edge of the boundary layer. The main idea is to always preserve the RANS mode inside the boundary layers. To achieve this, Menter and Kuntz (2002) introduced a variable which is the ratio of the model-related length scale (e.g. $k^{1/2}/\omega$) to the distance to the wall.

To force DES to preserve the RANS mode throughout the boundary layer, Spalart et al. (2006), following the ideas of Menter and Kurtz (2002), proposed several modifications to the original DES model. In one-equation models, such as the SA model, the model-related length scale has to be expressed through the eddy viscosity as $\sqrt{\frac{\nu_t}{S}}$ resulting in the parameter r in Equation (7.8), which is the square of the model length scale to the wall distance. The original parameter r (last formula in Eqn. 7.8) is replaced by a new parameter, r_d, in the delayed version of the SA DES model.

$$r_d = \frac{\nu + \nu_t}{\kappa^2 d^2 \cdot \max\left(\sqrt{\dfrac{\partial \overline{u}_i}{\partial x_j}\dfrac{\partial \overline{u}_i}{\partial x_j}}; 10^{-10}\right)} \tag{7.21}$$

This parameter is equal to one in the logarithmic layer, and falls gradually to zero towards the edge of the boundary layer. The new variable r_d is used to define a smooth function which has the value of one in the LES region, where $r_d \ll 1$, and zero elsewhere:

$$f_d = 1 - \tanh\left(\left[8r_d\right]^3\right) \tag{7.22}$$

Then, the model length scale, d_{DDES}, is redefined as follows:

$$d_{DDES} = d_{RANS} - f_d \max(0, d_{RANS} - d_{LES}) \tag{7.23}$$

where d_{LES} is given by Equation (7.11) This redefinition of the model length scale via the introduction of f_d prevents a too early switch to the LES mode. Setting f_d equal to 0 yields RANS ($d_{DDES} = d_{RANS}$), while setting f_d equal to 1 gives the regular DES formulation for the length scale ($d_{DDES} = \min(d_{RANS}, d_{LES})$). Thus, f_d forces DES to solve attached boundary layers in the RANS mode, independent of the grid resolution. As the end effect is to prevent DES from a premature switch to LES mode, the new version of DES was called Delayed DES (DDES). Though DDES was originally proposed as an improvement of the original SA based version of DES, DDES formulations of DES using other base RANS models are possible. One drawback of DDES is that the length scale d_{DDES} is now dependent on the solution via r_d and thus on time. Consequently, in many flows the mean solution becomes sensitive on the initial conditions (Fröhlich and von Terzi, 2008), which is an unwanted feature of DDES.

Improved Delayed DES (IDDES)

Shur et al. (2008) developed a further extension of the DES method which combines DDES with a special version of the two-layer model type described in section 7.2. The combined model is called Improved DDES (IDDES) and aims at being applicable also to attached boundary layer flows, including channel flows, thereby resolving the log-layer mismatch between RANS and LES regions in such flows. An important element is the new definition of the subgrid length scale, which does not only depend on the grid spacing but also on the wall distance. This allows increasing the resolved turbulence activity and together with a blending of the RANS and LES length scales adjusts the resolved log layer to the modelled one. In situations where the inflow has turbulent content and the grid is fine enough to resolve most of the turbulence except very close to the wall, IDDES reduces to the underlying two-layer model. In cases without inflow turbulent content, the IDDES model reduces to DDES, i.e. it gives a pure RANS solution for attached flows and a DES-like solution for massively separated flow. The length-scale relations, switching and blending functions for the case the Spalart-Allmaras is used as the base RANS model are rather complex and can be found in Shur et al. (2008). k-ω SST model based formulations of DDES and IDDES can be found in Gritskevich et al. (2012).

Shur et al. (2008) tested IDDES successfully in pure two-layer model applications (high Reynolds number turbulent channel flow, flow over a hydrofoil with trailing edge separation), in a natural DDES application to massively separated flow (flow past an airfoil at high angle of attack) and in complex flows where both branches of the model are active (flow past a backward-facing step). Additional promising IDDES applications are presented in Mockett and Thiele (2007). One potential problem is the possibility that the final solution depends on the initial conditions. For example, in a periodic channel flow simulation, depending on the level and type of initial fluctuations, IDDES can converge to either a steady RANS solution or to a statistically steady LES solution.

Modified Δ-based length scales

An issue in DES which is grid related is the definition of the local grid spacing, Δ, in grid cells with a high aspect ratio situated away from the walls (see also Breuer et al., 2003, Chauvet et al., 2007). A typical example is the case of a turbulent jet issuing from a nozzle. To resolve the attached boundary layer on the nozzle wall, the grid has to be very fine in the wall normal direction. In principle, a fast transition to LES mode is desirable downstream of the nozzle to allow development of the Kelvin Helmholtz instabilities in the ensuing mixing layers. The usual definition of Δ used in DES (($\Delta = \max(\Delta_x, \Delta_y, \Delta_z)$) does not allow a sufficiently fast development of these instabilities in regions where high aspect ratio cells are present. One possibility is to define $\Delta = (\Delta_x \Delta_y \Delta_z)^{1/3}$ as in classical LES. Chauvet et al. (2007) proposed using $\Delta = (n_x^2 \Delta_y \Delta_z + n_y^2 \Delta_z \Delta_x + n_z^2 \Delta_x \Delta_y)^{1/2}$ for cases when the vorticity is closely aligned with one of the grid lines, where (n_x, n_y, n_z) is the unit vector aligned with the vorticity.

7.5 SCALE-ADAPTIVE SIMULATION (SAS) MODEL

This model, proposed first by Menter and Egorov (2005) and described best in Menter and Egorov (2010), employs a special RANS model that in flows with natural unsteadiness resolves in a 3D unsteady calculation turbulent fluctuations and hence has LES-like behaviour. The RANS model introduces the second derivative of the velocity in the source term of the length-scale determining model equation and thereby the von Karman length as a second length scale. This term reduces the eddy viscosity in flows with unsteadiness to such an extent that large scales are broken up into smaller ones, reducing further the eddy viscosity until the grid limit is reached. Thus, depending on the grid spacing, time step and numerical dissipation, a substantial part of the turbulent spectrum can be resolved. As the scales cannot become smaller than the mesh size, and hence related motions cannot be resolved, their (dissipative) effect must be accounted for by a SGS-type viscosity related to the mesh size – hence the model switches into LES mode when the RANS eddy viscosity becomes smaller than the SGS viscosity. The SAS model is therefore clearly a hybrid RANS-LES model. It behaves in many situations similar to DES, but has a safer transition from RANS to scale-resolving behaviour on ambiguous grids, a problem with DES discussed in section 7.4.4.

The SAS model is based on the k-kL two-equation RANS model of Rotta (1972), employing model equations for k and the product kL, where L is the integral length scale of the turbulence. The particularity of the k-kL model is the presence of a second length scale in the source terms of the length-scale determining kL equation. This second length scale involves the second derivative of the velocity field. In the SAS model, the original transport equation for kL is replaced by a transport equation for $\Phi = \sqrt{kL}$. This quantity is proportional to the eddy viscosity. The second length scale is, in fact, the von Karman length scale L_{vk} defined below in Equation (7.26). The equations of the RANS model are summarized below:

$$\frac{\partial k}{\partial t} + \bar{u}_j \frac{\partial k}{\partial \bar{x}_j} - \frac{\partial}{\partial x_j}\left(\frac{\hat{v}_t}{\sigma_k}\frac{\partial k}{\partial x_j}\right) = G - c_\mu^{3/4}\frac{k^2}{\Phi} \tag{7.24}$$

$$\frac{\partial \Phi}{\partial t} + \bar{u}_j \frac{\partial \Phi}{\partial \bar{x}_j} - \frac{\partial}{\partial x_j}\left(\frac{\hat{v}_t}{\sigma_\Phi}\frac{\partial \Phi}{\partial x_j}\right) = \frac{\Phi}{k}G\left(\varsigma_1 - \varsigma_2\left(\frac{L}{L_{vk}}\right)^2\right) - \varsigma_3 k \tag{7.25}$$

$$\hat{v}_t = c_\mu^{1/4}\Phi$$

$$L = \frac{\Phi}{\sqrt{k}}; \quad L_{vk} = \kappa\left|\frac{U'}{U''}\right|; \quad U'' = \sqrt{\frac{\partial^2 \bar{u}_i}{\partial x_k \partial x_k}\frac{\partial^2 \bar{u}_i}{\partial x_j \partial x_j}}; \quad U' = \sqrt{2\bar{S}_{ij}\bar{S}_{ij}} \tag{7.26}$$

where the expression of the production term, G, is given by Equation (7.14) and the model constants are $c_\mu = 0.09$, $\varsigma_1 = 0.8$, $\varsigma_2 = 1.47$, $\varsigma_3 = 0.0288$, $\sigma_k = \sigma_\Phi = 2/3$. The term containing the second derivative of the velocity field (U'' in Eqn. 7.26) is used to define the von Karman length scale L_{vk}.

The presence of the term with $\left(\frac{L}{L_{vk}}\right)^2$ in the production term in the Φ-Equation (7.25) yields a lower eddy viscosity compared to models that do not involve this term. This is because the von Karman length scale adjusts to the smallest resolved scales until the grid limit is reached (hence the name Scale Adaptive Simulation). This lower eddy viscosity allows fluctuations to arise and to be sustained in the simulation under unstable flow conditions, similar to a LES simulation but using different modelling mechanisms and without an explicit influence of grid spacing in this RANS-model simulation. Of course, fluctuations can be sustained only for frequencies that are sufficiently resolved by the grid spacing and time step. However, the eddy viscosity given by the model should not decay below the level given by a SGS model in the regions where the model is in unstable mode. Otherwise, there will be a pile-up of energy around the cut-off wavenumber of the grid which will result in numerical oscillations. This can happen as the model, having no information on the cut-off limit, provides an eddy viscosity small enough to allow further cascading to scales smaller than the grid limit.

To provide sufficient damping of the smallest resolved scales at the high wave number end of the spectrum, an LES component was added to later versions of the model as described in Menter and Egorov (2010). This was done by introducing an eddy-viscosity limiter that chooses the eddy viscosity in SAS as the maximum between $\hat{\nu}_t$ from the RANS model and the one given by a SGS model. Such a limiter should not impact the simulation when the SAS model is in steady RANS mode. A dynamic Smagorinsky SGS model would be an obvious choice. However, the WALE model discussed in section 3.3.2 was preferred due to its robustness and easy implementation. Thus, the final expression of the eddy viscosity in SAS is:

$$\nu_t = \max(\hat{\nu}_t, \nu_t^{WALE}) \tag{7.27}$$

where ν_t^{WALE} is the SGS eddy-viscosity given by Equation (3.23). The limiter turns SAS into a LES when the time step is small enough (Courant number is less than one) to allow resolution of the fluctuations with scales close to the grid size. When the time step is larger, the limiter is not active and SAS is in the RANS (or rather URANS) mode, with the eddy viscosity adjusting to the resolved scales in the flow which are the larger the larger the time step is. For a large enough time step, a fallback steady RANS solution results.

The above model can also be converted into an equivalent k-ε, k-ω or SST model. The SST based version of SAS is the most popular one. The resulting transport equations for ε and ω contain an additional source term (e.g., see Egorov and Menter, 2008, Menter and Egorov, 2010). Several other minor changes are needed in the transport equations to preserve the RANS behaviour of the SST model for boundary layer flows (see Menter and Egorov, 2005). One-equation versions based only on the transport equation for Φ were also proposed by Menter et al. (2006). In the one-equation version, the turbulent kinetic energy is estimated as $k = c_\mu^{-1/4} U'' \Phi$. The numerical discretization techniques recommended for SAS calculations are similar to those for DES. The SAS model was incorporated into various codes, including commercial CFD codes, and has been successfully tested for a wide range of problems (Egorov et al. 2010). Though most of these tests concerned applications in mechanical and aerospace engineering, the model is expected to perform well in applications relevant to hydraulics.

7.6   FINAL COMMENTS ON HYBRID RANS-LES MODELS AND FUTURE TRENDS

The chapter presented what the authors believe are the most popular approaches and those that incorporate, at a conceptual level, the right physics to simulate high Reynolds number flows in complex geometries using hybrid RANS-LES methods. The material presented here is by no means complete – and more emphasis was put on models that were already applied or have the features which make them likely to be applied for flows and transport processes of interest in hydraulics. Most probably, the final success of these approaches will depend on their robustness, on how sensitive the results are to the numerical discretization scheme and to the grid, and on how easy they can be implemented in multi-block solvers using structured and unstructured grids.

A general problem with hybrid RANS-LES approaches employing also in the LES component a RANS-based model is that the SGS model is constrained because of calibration in the RANS mode. Such SGS models obtained by a redefinition of the turbulent length scale in a base RANS model incorporate less physics compared to approaches developed specifically for LES (e.g., dynamic Smagorinsky model or models using a transport equation for the SGS kinetic energy). As a result, it is probably impossible such hybrid models will accurately predict all types of flows. The recently proposed IDDES version of DES is trying to enlarge the range of flows that can be accurately predicted using such methods and to make the model predictions less sensitive to the gridding. It is expected that IDDES will be further refined in the future and should then be the preferred version of DES for general applications. More complex RANS closures (e.g., algebraic-stress models were used in the hybrid RANS-LES method proposed by Abe, 2005) can be used to account for the turbulence anisotropy in the RANS sub-domains. The use of SAS models for hybrid RANS-LES calculations shows considerable promise. A main limitation of the present versions of SAS is the fact that unsteadiness cannot be enforced for flows for which the model produces a steady solution. Ongoing work in this area concerns modifications that will allow enforcement of unsteady behaviour in such cases (see Menter et al., 2009).

The TBL-based two-layer model of Wang and Moin (2002) can be further refined to incorporate more advanced RANS closures that use transport equations for the turbulence quantities rather than a mixing-length model to estimate the eddy viscosity. Diurno et al. (2001) already used the SA model to predict the eddy viscosity in the RANS subdomain of a simpler two-layer model. On the other hand, the implementation of the most successful type of models that solve the TBL equations on a separate embedded grid to resolve the near-wall flow is quite complicated in an unstructured grid environment, and the model predictions are quite sensitive to the way the coupling is achieved at the interface. Whether or not such models will become popular in hydraulics, where multiple wall surfaces of complex shapes are generally present, is an open question.

Finally, the capability of hybrid RANS-LES methods to accurately simulate scalar transport is very important for many applications in hydraulics and river engineering. For hybrid RANS-LES methods in which the base model is a RANS model, the modeling of the eddy diffusivity is, in most cases, done very crudely (e.g., the eddy diffusivity is assumed to be proportional to the eddy viscosity). Accounting for stratification effects, which are important in many applications in hydraulics and river engineering, is still a considerable challenge for RANS modeling. Methods that use SGS models specifically developed for LES and a dynamic approach for calculating the model coefficients in the LES sub-domains do not face this challenge and have a clear advantage. For example, the dynamic Smagorinsky model was successfully used to predict scalar transport in non-stratified and stratified environments for a wide range of turbulent flows.

Chapter 8

Eduction of turbulence structures

In large-eddy simulations the energy-containing motions are calculated directly, which allows an explicit simulation of the unsteady, large-scale turbulence structures in the flow. Large-scale turbulence structures, also called "coherent structures" or "quasi-coherent structures", have been the subject of many experimental studies (e.g. pioneering work by Kline et al., 1967; Corino and Brodkey, 1969; Grass, 1971; Willmarth and Lu, 1972) and have been identified in these early works on boundary-layer turbulence mainly through visualization or conditional sampling techniques. Though no clear definition of "coherent structures" exists, it is commonly agreed that a coherent structure is a region of space (or time) within which the flow field has a characteristic coherent flow pattern (Pope, 2000). In his review article, Robinson (1991) provides the following definition: *"a three-dimensional region of the flow over which at least one flow fundamental variable exhibits significant correlation with itself or with another variable over a range of space and/or time that is significantly larger than the smallest scale of the flow"*. Nezu and Nakagawa (1993) introduce and discuss coherent structures in the context of hydraulic engineering open-channel flows and they distinguish between "bursting-structures" near the channel bed and "large-scale vortical structures" that are generated by the mean flow and/or the channel geometry away from the bed. In Table 8.1, which builds upon the categorization according to Robinson (1991), the most common coherent structures occurring in open-channel flow or in flows of hydraulic engineering interest are listed. The first five structures are considered "bursting-structures", whereas the last seven are categorized as "large-scale vortical structures".

As table 8.1 shows, a large number of different types of coherent structures exists, each of them caused by different physical mechanisms. However, it was not before the advent of supercomputers in the early 1980's, which enabled direct numerical simulations of low Reynolds number channel flow, that many of the three-dimensional coherent structures could be educed and quantified. Since then, coherent structures in the flow and their generation mechanisms were identified from instantaneous three-dimensional pressure, velocity and vorticity fields available from both direct numerical simulations and large-eddy simulations. In recent years, a good number of large-eddy simulations have been performed to complement existing laboratory experiments. Often, the data from an experiment are used to validate the simulation, which in turn then provides all flow quantities at every instant in time and at every location in the flow. This is particularly significant for quantities that are very difficult

Table 8.1 Coherent structures and associated events found in turbulent flows of hydraulic interest.

Structure	Description
Low and high speed streaks	Elongated areas of alternating high and low speed fluid
Bursting	Lift up and bursting of low speed streaks including sweeps and ejections
Ejection-events	Ejection of low-momentum fluid away from the wall
Sweep-events	Inrush of high-momentum fluid from outer layer
Hairpin Vortices	Near-wall spanwise vortices with elongated trailing legs
Horizontal/Vertical Rollers	Vortices with distinct rotation about a one-dimensional axis
Horseshoe/Necklace Vortices	Vortices as a result of three-dimensional boundary layer separation due to flow obstruction
Von Karman type vortices	Shed vortices in the wake behind bluff bodies
Kelvin-Helmholtz billows	Internal instabilities as a result of strong velocity shear
Tornado vortices	Funnel-shaped vertical axis vortices
Shear layer backs	Sloping streamwise velocity discontinuities
Boils, Kolk-boils	Strong upward vortex motion initiated near the bed

to measure, e.g. pressure fluctuations, or at locations in the flow where measurements are impossible, e.g. very close to the wall.

Generally, large-eddy simulations provide an enormous amount of high-resolution data and by employment of adequate coherent structure eduction techniques insight into the physical mechanisms can be obtained. Depending on the structure to be educed, different eduction techniques are available to analyze and visualize the flow. The most prominent ones will be introduced in the following.

8.1 STRUCTURE EDUCTION FROM POINT SIGNALS: TWO-POINT CORRELATIONS AND VELOCITY SPECTRA

Using velocity or pressure fluctuation signals in space and time to educe, or rather detect and quantify, coherent structures can be done in a way analogous to that in experiments, i.e. through conditional sampling. The technique of conditional sampling and subsequent averaging can be seen as a method for quantifying flow coherence in a turbulent flow. In general, conditional sampling pertains to "averaging under certain conditions", in which detection functions are used to define the type of averaging (Antonia, 1981). A detailed description of conditional-sampling techniques for open-channel turbulence can be found in Nezu and Nakagawa (1993). The advantage of conditional sampling within the framework of LES is that velocity and pressure signals can be sampled simultaneously at any point in the domain and at very high temporal and/or spatial resolution. Of the many other possible ways to educe structures from time signals, two will be introduced and discussed in the following.

Two-point correlations are used as a statistical tool in turbulence research to determine length scales of dominant structures. The normalized correlation coefficient tensor, R_{ij}, (obtained at a location x, z from velocity signals at two points separated in the y-direction) is defined as:

$$R_{ij}(r) = \frac{\langle u_i(y,t) \cdot u_j(y+r,t) \rangle}{\langle u_i(y,t) \cdot u_j(y,t) \rangle} \tag{8.1}$$

where r is the distance between sampling points. Figure 8.1a sketches a typical two-point velocity function in a turbulent flow. For zero separation ($r = 0$), $R(r)$ is of course 1.0, and after a certain distance, $R(r)$ goes to zero, i.e. the signal becomes uncorrelated. The integral length scale $L = \int_{r=0}^{\infty} R(r)\,dr$ can be determined by integration of the function. The distance over which $R(r)$ falls to zero provides another important turbulent length-scale, i.e. it gives the approximate size of a large eddy. Figure 8.1b presents the two-point correlations of the streamwise, spanwise and wall-normal velocity fluctuations, R_{uu}, R_{vv} and R_{ww}, in the spanwise direction in developed channel flow over a smooth bed from the DNS of Kim et al. (1987). A corresponding snapshot of the instantaneous streamwise velocity fluctuation distribution in a horizontal near-wall plane is given in Figure 8.6. From this it can be seen that the near-wall flow is dominated by low- and high-speed streaks; however this instantaneous contour plot only provides a qualitative picture of the flow, while the two-point correlations of a sampled time signal along the spanwise direction of the flow provide quantitative information of the low- and high-speed streaks. These streaks are a result of counter-rotating near-wall vortex pairs with size σ (as sketched in Fig. 8.2), which entrain high-momentum fluid into a high-speed streak and low momentum fluid into a low-speed streak. The two-point correlation of the streamwise velocity fluctuation, R_{uu}, allows quantifying the streak spacing, λ, and the two-point correlation of the wall normal velocity fluctuation, R_{ww}, allows the determination of the average vortex roller size (Rajaee et al., 1995). In Figure 8.2b R_{uu}, reaches a minimum at a distance $y^+ = 50$, suggesting an average streak spacing of $\lambda^+ = 100$ and the vortex roller size is found to be $\sigma = 30$ wall units.

Velocity spectra can be employed to detect the occurrence of turbulence structures that are regularly repeating in time. As an example, shed vortices in a separated flow can be substantiated through spectra of the three fluctuating velocity components.

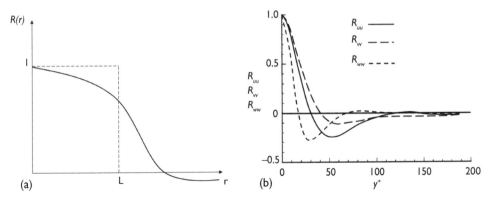

Figure 8.1 a) Sketch of a typical two-point correlations (b) two-point correlation of the streamwise, spanwise and wall normal velocity fluctuations, R_{uu}, R_{vv} and R_{ww} in the spanwise direction in channel flow (adapted from Kim et al., 1987).

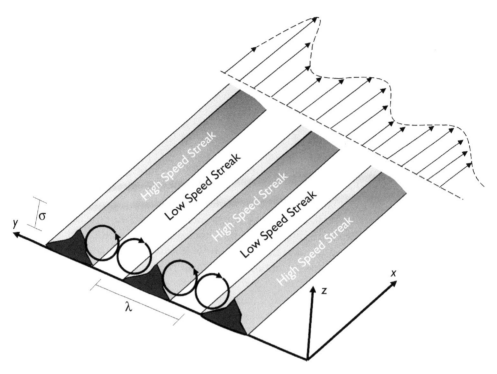

Figure 8.2 Sketch of near wall turbulence structures.

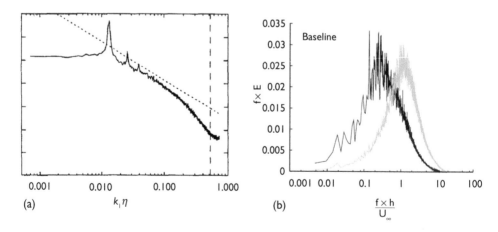

Figure 8.3 (a) Velocity spectrum of spanwise velocity fluctuation at a location in the wake behind a long cylinder at $Re_D = 3900$ in a log-log plot; the frequency is normalized with the Kolmogorov length scale and the dashed line represents the $k^{-5/3}$ line (b) pre-multiplied velocity spectra in the flow over a hump at two locations downstream of the hump in a semi-logarithmic plot. (a) from Ong and Wallace (1996) and (b) from Avdis et al. (2009), both reproduced with kind permission from Springer Science + Business Media B.V.

A velocity spectrum is obtained through fast Fourier transform of the signal. Shedding frequencies can be quantified by identification of a distinct peak in the velocity spectra. For instance, for the flow around a long circular cylinder, the spectrum of the spanwise velocity exhibits a peak. Figure 8.3a presents such a spectrum (*pds*) of the spanwise velocity fluctuation, and a distinct peak is observed at a dimensionless frequency (i.e. the Strouhal number, St) of 0.2, at which vortex shedding occurs. Another example of how velocity spectra are interpreted with regard to coherent structures is provided in Figure 8.3(b). There, spectra are plotted as a function of the Strouhal number, St, at two locations in the flow over a hump (Avdis et al., 2009). The plot exhibits the shift of the spectrum peak from St = 1.0 at a location close to the hump, to St = 0.2 in the shear layer further away from the hump.

8.2   STRUCTURE EDUCTION FROM INSTANTANEOUS QUANTITIES IN 2D PLANES

A number of turbulent flow structures, e.g. von Karman vortices shed from a long circular cylinder, can be visualized by contours of instantaneous velocity (components), contours of vorticity, streaklines, velocity vectors, or vectors of velocity fluctuations. For the flow over a long cylinder, Figure 8.4 presents contours of instantaneous streamwise velocity (a), velocity vectors and streamlines at an instant in time in a blow-up close to the cylinder (b), and streamlines in a frame moving with the average vortex velocity (c) in a 2D plane. The vortex shedding can be discerned from the velocity contours, a more quantitative picture is obtained from instantaneous streamlines, from which the position of each vortex can be determined. An alternative to using instantaneous velocity contours is the use of contours of components of the vorticity or the vorticity magnitude from which vortices and their direction of rotation can be identified (Fig. 8.5).

Vorticity, on the other hand, is not always an unambiguous quantity, because an area of high shear would also give a high value of vorticity and does not necessarily feature a vortex (e.g. Jeong and Hussain, 1995). This can be also seen in Figure 8.5, where there are areas close to the cylinder in which the vorticity is high but no vortices are present. Kline and Robinson (1989) propose to define a vortex as a region of high vorticity around which a pattern of streamlines is roughly circular when viewed in a frame moving with the vortex. Adrian et al. (2000) applied Galilean decomposition, i.e. decomposition of the velocity field into a convection velocity plus the deviations therefrom, and educed vortices through the deviation vectors, which are the vectors seen in a frame of reference moving at the convection velocity. Galilean decomposition is often used to analyze 2D PIV (Particle Image Velocimetry) data, however the selection of an appropriate convective velocity is difficult because vortices can travel at different speeds depending on their position in the flow. For instance, Hinterberger et al. (2008) and Stoesser et al. (2008) have chosen the bulk velocity as convective velocity. Similarly, Bomminayuni and Stoesser (2011) superimposed vectors of the instantaneous velocity fluctuation onto contours of vorticity in order to identify instantaneous vortices. This is demonstrated in Figure 9.10 in the Applications chapter 9.2 for flow over a rough bed composed of hemispheres. A clear distinction between individual vortices and areas of high shear can be obtained. For instance, in Figure 9.10(a) areas of high positive spanwise vorticity are found over the hemisphere suggesting clockwise

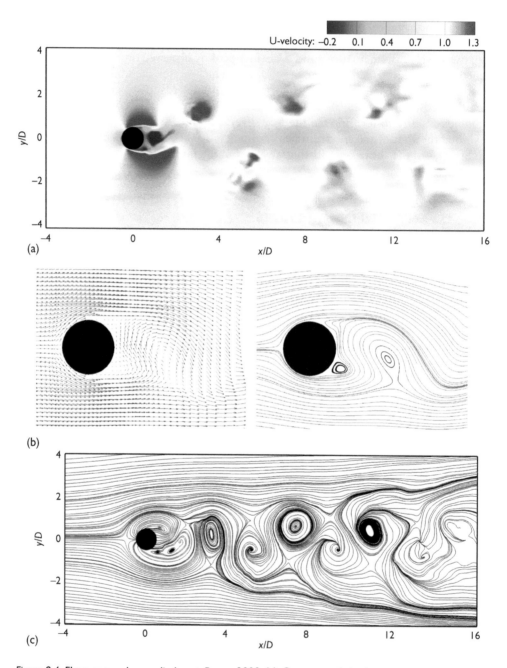

Figure 8.4 Flow over a long cylinder at $Re_D = 3900$ (a) Contours of the instantaneous streamwise velocity in a 2D plane (b) vectors and streamlines of the instantaneous flow close to the cylinder and (c) instantaneous streamlines in a frame moving with the average vortex velocity. Simulations by Stoesser (unpublished results).

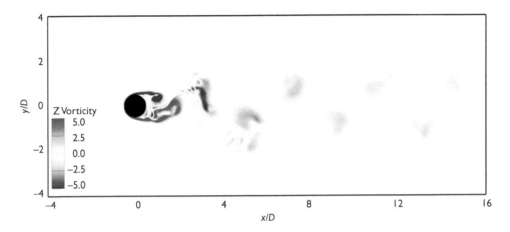

Figure 8.5 Contours of the vorticity of flow over a long cylinder at $Re_D = 3900$ in a 2D plane. Simulations by Stoesser (unpublished results).

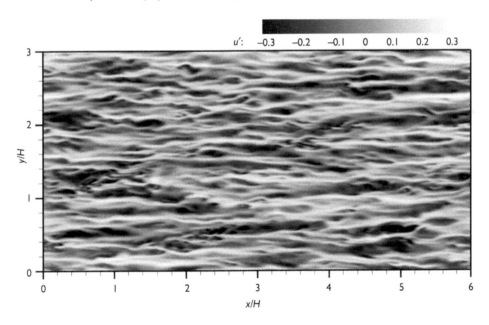

Figure 8.6 Chanel flow: Contours of instantaneous streamwise velocity fluctuation (u') in a 2D plane approximately ten wall units above a smooth wall. Simulations by Stoesser (unpublished results).

rotating vortices, but as Figure 9.10b reveals, these areas of high vorticity do not necessarily feature vortices but are rather areas of high shear. Another usage of velocity perturbation vectors in an instantaneous flow is to identify sweep and ejection events. As Figure 9.10b reveals, at the selected instant in time, several sweep (S) and ejection (E) events occur.

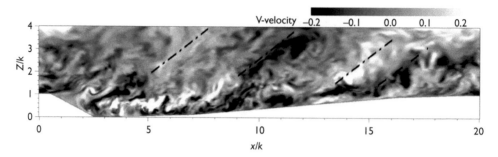

Figure 8.7 Contours of instantaneous spanwise velocity of the flow over a fixed sand dune in a longitudinal 2D plane. Shear layer backs are indicated by the dashed lines. From Stoesser et al (2008), reproduced with permission from ASCE.

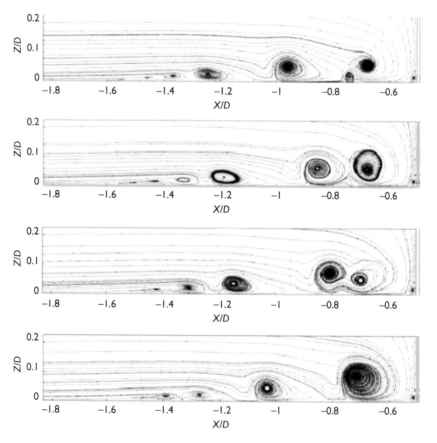

Figure 8.8 Streaklines at various instants in time in a 2D longitudinal plane of the near wall flow upstream of a wall mounted cylinder. From Escauriaza and Sotiropoulos (2011), reproduced with kind permission from Springer Science + Business Media B.V.

Other structures that can be educed through contours of various quantities in 2D planes include high- and low-speed streaks, shear-layer backs, horseshoe vortices or hairpin vortices amongst others. Figure 8.2 sketches idealized high- and low-speed streaks in the flow over a wall and Figure 8.6 shows contours of instantaneous streamwise velocity fluctuations in a 2D plane approximately ten wall units above a smooth wall. The presence of streaks is clearly visible. Shear layer backs are inclined regions of streamwise velocity discontinuities, which can occur as a result of near-wall turbulence structures being lifted up and convected towards the water surface, so that the flow is continuously disrupted in the streamwise direction. Shear-layer backs can be educed best by contours of the spanwise velocity in a longitudinal 2D plane (Figure 8.7). The eduction of a horseshoe vortex system, e.g. around a single cylinder placed on a wall, with streamlines of the instantaneous velocity in a longitudinal 2D plane (here through the cylinder axis) is illustrated in Figure 8.8. Therein, the near-bed flow in front of the cylinder is visualized, and the streamlines reveal the occurrence of the complex vortex system. The identification of hairpin vortices, or hairpin packets, can be achieved by plotting the swirl strength in a longitudinal 2D plane. The swirl strength criterion is computed using a reduced tensor that contains the in-plane velocity gradients only (Zhou et al., 1996, Adrian et al., 2000). This is basically the 2D version of the λ_2-criterion introduced further below and has been used frequently to analyze 2D PIV measured velocity data. Contours of the two-dimensional swirl strength isolate regions that are swirling about the axis normal to the plane. Figure 9.10(c) in the Applications chapter 9.2 presents contours of the swirl strength together with perturbation vectors for flow over a rough bed. Areas of fluid rotation at an instant in time are highlighted by the grey contours.

8.3 STRUCTURE EDUCTION FROM ISOSURFACES OF INSTANTANEOUS QUANTITIES IN 3D SPACE

The advantage of large-eddy simulations over the most advanced experimental techniques is that all three velocity components as well as the pressure are available at any instant in time. Some coherent structures are characterized by vortices rotating about an axis, such as spanwise rollers or von-Karman type vortices. Such vortices usually feature a local pressure minimum in their core and can be visualized through iso-surfaces of the instantaneous pressure fluctuation $p' = (p - \langle p \rangle)/(\rho U_\infty^2)$. The selection of an appropriate (negative) value of p' is rather subjective and there are no unambiguous rules. However, using pressure fluctuations, Stoesser et al (2008) educed spanwise rollers and other structures occurring in the flow over a two-dimensional fixed dune (Fig. 8.9). An isosurface of $p'/(\rho U_\infty^2) = -0.08$ provided a reasonably lucid picture of the turbulence structures i.e. spanwise rollers (denoted R1 and R2), kolk-boil vortices in the form of hairpins (HP1 and HP2) and elongated streamwise vortices (E). The value of p' was chosen somewhat arbitrarily, however values greater or smaller than 0.08 would either include too many structures and overcrowd the image or not reveal all of the structures seen in Figure 8.9.

Obviously, a preselected isosurface-value of the pressure fluctuation is not an objective eduction technique, and finding the "right" value involves a bit of trial-and-error. Also, one threshold may not identify all structures in the flow. Moreover, isosurfaces of pressure fluctuations may not educe a coherent structure at all, regardless of the value

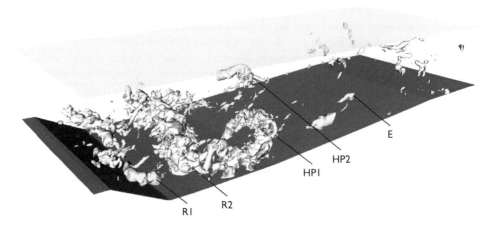

Figure 8.9 Isosurfaces of the pressure perturbation $p' = 0.08$ at an instant in time of the flow over a two-dimensional fixed dune. From Stoesser et al, (2008),reproduced with permission from ASCE.

chosen, because local pressure minima can be found in a flow that has no vorticity (Jeong and Hussain, 1995). An alternative is to use isosurfaces of vorticity (magnitude) but then again, areas of significant shear (e.g. near a wall) also exhibit local maxima of vorticity without the presence of vortices. Figure 8.10 presents visualized von Karman vortex shedding behind a cylinder by isosurfaces of pressure fluctuations (a), and isosurfaces of the vorticity magnitude (b). Obviously, both techniques provide a qualitative impression of the flow. Using an isosurface of a certain pressure fluctuation, the structures appear to be smoother but the prevailing ribs and rolls and other secondary structures are not visualized. On the other hand, vorticity magnitude educes these secondary structures but also visualize the shear layer from which the vortices evolve.

In recent years, more robust definitions of vortices have been proposed (Hunt et al., 1988, Jeong and Hussain, 1995), aiming at removing some of the ambiguity involved with pressure minima and vorticity. These methods are known as Galilean invariant vortex identification techniques, which imply that the magnitude of the convective velocity does not play a role and represent a vortex criterion that remains invariant under coordinate changes. Probably the two most common vortex identification techniques are due to Hunt et al. (1988), and Jeong and Hussain (1995); both use the following velocity gradient decomposition:

$$\nabla \mathbf{v} = S_{ij} + \Omega_{ij} \tag{8.2}$$

where $S_{ij} = \frac{1}{2}\left(\frac{\partial u_i}{\partial x_j} + \frac{\partial u_j}{\partial x_i}\right)$ is the rate of strain tensor and $\Omega_{ij} = \frac{1}{2}\left(\frac{\partial u_i}{\partial x_j} - \frac{\partial u_j}{\partial x_i}\right)$ is the vorticity tensor, which are the symmetric and antisymmetric components of $\nabla \mathbf{v}$. Hunt et al. defined a vortex as a region with positive second invariant of $\nabla \mathbf{v}$ i.e. $Q > 0$, in which the pressure attains a local minimum. Q is defined as follows:

$$Q \equiv \frac{1}{2}\left(\left|\Omega_{ij}\right|^2 - \left|S_{ij}\right|^2\right) \tag{8.3}$$

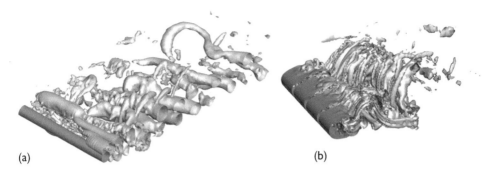

(a)   (b)

Figure 8.10 Isosurfaces of the pressure fluctuation $p' = 0.5$ (a) and vorticity magnitude $|\omega| = 2$ (b) at an instant in time in the flow over a long cylinder. Simulations by Stoesser (unpublished results).

and hence as a local region in which vorticity dominates shear strain, which allows to distinguish between rotation and shear. However, a region with $Q > 0$ does not necessarily imply a local pressure minimum. The second Galilean-invariant vortex-identification method is according to Jeong and Hussain (1995), and is known as the λ_2-criterion. Jeong and Hussain (1995) define vortices as regions with:

$$\lambda_2 \left(\Omega_{ij}^2 + S_{ij}^2 \right) < 0 \tag{8.4}$$

in which $\lambda_2(A)$ denotes the second eigenvalue of a symmetric tensor A.

With the two definitions, vortices in a flow can be educed more or less reliably, and in many cases Q- and λ_2-criteria produce almost identical results. For instance, Figure 8.11 presents turbulent structures in the flow over a wall-mounted hemisphere. The left hand side of Figure 8.11 is the eduction of the structures using the λ_2-criterion and the right hand side shows the structures as educed by employing the Q-criterion. There is virtually no visible difference and both techniques provide good impressions of the prevailing structures.

Figure 8.12a presents snapshots of the velocity-colored iso-surface of $\lambda_2 = 3V_0/D$ (where V_0 is free stream velocity and D the cylinder diameter) of the flow around two cylinders in tandem (from Garbaruk et al 2012). Large-scale vortex shedding as well as secondary vortex structures can be identified. As yet another example of the Q-criterion, Muld et al. (2011) present isosurfaces of Q colored by the instantaneous pressure for the flow over a surface-mounted cube, which is presented in Figure 8.12(b). The horseshoe vortex upstream of the cube and the vortices that are a result of flow separation from the leading edges of the cube are educed evidently.

It should be added that the two popular definitions given in Equations 8.3 and 8.4 provide a robust framework for the identification of a(ny) vortex in the flow. Unfortunately, this does not necessarily make the task of educing coherent structures in a turbulent flow any easier than using a certain pressure fluctuation value. In fact, high Reynolds number flows in hydraulic engineering feature an uncountable amount of vortices of different size and of different strength. Any of the two criteria could,

Figure 8.11 Coherent structures in the flow over a hemisphere educed by the − (left) and Q-criteria (right). Simulations by Bomminayuni and Stoesser (unpublished results).

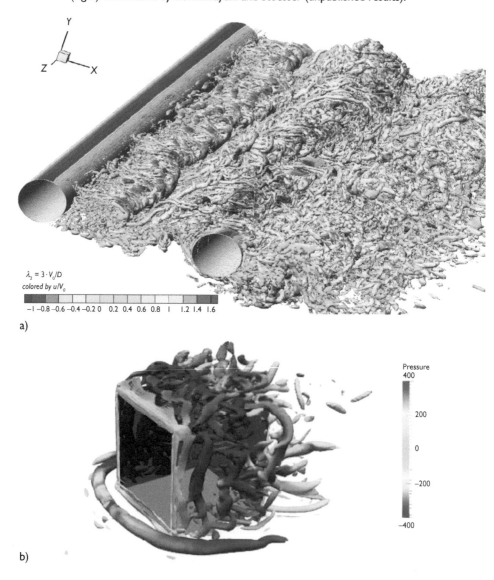

Figure 8.12 Isosurfaces of the Q-criterion at an instant in time; (a) of the flow around two cylinders in tandem (from Garbaruk et al 2012) and (b) of the flow over a wall mounted cube (from Muld et al., 2011, reproduced with kind permission from Springer Science + Business Media B.V.).

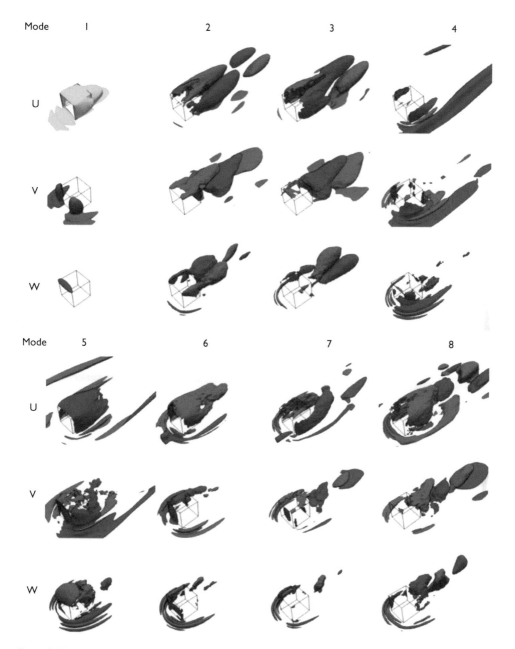

Figure 8.13 Isosurfaces of positive(red) and negative (blue) velocity components of the first four POD modes in flow over a surface mounted cube (adapted from Muld et al., 2011, with kind permission from Springer Science + Business Media B.V.).

in theory, identify these vortices but to educe the most energetic ones, selection of a proper threshold is required, which again can lead to a trial and error procedure resulting in more or less subjectivity. Additionally, most often one large vortex is surrounded by many small vortices (as a result of shear and rotation), and isolating the large structure is not an easy task. The fact that the velocity field of flows in hydraulics computed by LES is non-smooth, for instance due to localized high velocity gradients (or even due to numerical instabilities or oscillations), may result in many small artificial vortices that overcrowd the visualization and obscure the relevant structures.

Finally, Proper Orthogonal Decomposition (POD), a method first applied to study turbulence by Lumley (1967) should be introduced. POD, also known as Karhunen-Loeve decomposition, identifies the most energetic motions in the flow by decomposing a fluctuating velocity field into a set of selected basis functions and basis-function coefficients, and allows quantification of the contribution of coherent structures to the mean flow. An obvious choice of a basis function is the use of Fourier modes, which allows the calculation of the energy in each mode. POD is a commonly used technique for the identification of dominant energy-containing structures in the flow (e.g. Holmes et al. 1996). Snapshots of the instantaneous velocities are collected over a period of time and the time-averaged velocity is removed to obtain the fluctuating velocities. The application of the POD to simulations of turbulent channel flow was presented for instance by Moin and Moser (1989) who showed that the most organized structures of turbulence could be extracted and their contribution to kinetic energy and turbulence production could be quantified. An example of a POD-decomposed flow field is given in Figure 8.13 (from Muld et al., 2011), in which the most energetic contribution to the flow around a surface mounted cube is visualized by plotting isosurfaces of the three velocity components of the first eight, i.e. the most energetic, POD modes. Mode 1 represents the mean flow, for instance the green isosurface of the u-component encloses the reverse flow, i.e. it visualized flow separation and flow reattachment and the separation bubble downstream of the cube. The isosurfaces of the higher modes represent the unsteady and asymmetric features, here vortex shedding into the wake (e.g. modes 2 and 3 of the u-component). Modes two and three are closely connected and represent low frequency, most energetic, vortices rotating in clockwise and counterclockwise directions, respectively. Higher modes of the u-component exhibit structures of the diverse shear layers, or the presence of the horseshoe vortex (e.g. mode 6 in the u-velocity and w-velocity components). The interpretation of POD data plots is not an easy task, so that POD plots alone may not educe coherent structures, however POD plots can be used complementary to other techniques, like two-dimensional contour plots of vorticity or three-dimensional plots of isosurfaces of Q or λ_2.

Chapter 9

Application examples of LES in hydraulics

This chapter presents examples of applications of LES and Hybrid methods to a variety of hydraulic flows, many of them of environmental relevance. The examples start with simple developed, straight, smooth-bed open-channel flow, adding then complexities such as roughness, vegetation, curvature, stream merger, structures of various kinds and gravity effects. For each example, a short introduction to the flow situation is provided and available experimental studies are discussed, and then details on the simulations are presented, such as information on the calculation domain, the boundary conditions, the numerical grid and the specifics of the method used. As mentioned already in the Preface, there is clearly a bias towards calculations performed and results obtained by the authors because these are best known to them, but results of others are also included. Some comparisons of results for statistical quantities with experiments are provided (more can be found in the background literature), but the emphasis is placed on results for unsteady features and in particular turbulent structures of the flows in order to demonstrate the great potential of eddy-resolving methods such as LES for situations where these features play an important role. For some examples the superiority over RANS/URANS calculations is shown by including results of the latter for comparison.

9.1 DEVELOPED STRAIGHT OPEN CHANNEL FLOW

The first application example taken from Hinterberger et al. (2008) concerns the most basic hydraulic flow, namely developed flow in a straight open channel with smooth bed, here at a Reynolds number based on friction velocity u_* and channel depth h of $Re_\tau = 590$ (Re_h based on h and bulk velocity is 11,000). The channel is considered infinitely wide and the flow developed and hence periodic boundary conditions were applied in both streamwise and spanwise direction. The size of the computational domain was $8\pi h$ and $2\pi h$ in these directions, respectively. The LES was carried out with the standard Smagorinsky SGS model, but with near-wall damping of ν_t. The numerical grid had 8 Mio points and the wall-parallel resolution was in wall units $\Delta x^+ = 29$ and $\Delta y^+ = 14.5$ in streamwise and spanwise direction, respectively. The first grid point had a distance of $z_p^+ = 1.5$ from the bed. Hence, the conditions for a well-resolved LES were satisfied and no special near-wall treatment was necessary. In Figure 9.1 (left) calculated distributions over the depth of mean velocity and RMS values of the fluctuating components and of the shear stress $<u'w'>$ are compared with DNS results of Moser et al. (1999) for closed channel flow at the same Re_τ

(based on channel half width). The agreement can be seen to be good, but it should be noted that near $z/h = 1$ the fluctuations behave differently, as in the LES the vertical fluctuations are damped by the free-surface boundary while in the DNS this location is in the middle of the closed channel. The LES distributions agree also fairly well with measurements of Nezu and Rodi (1986). Figure 9.1 (right) gives evidence of the resolution of turbulent fluctuations by the LES. The figure shows for one instant in time the streamwise distribution of u-fluctuations at different depths z/h. As was to be expected from the RMS profile, the fluctuation amplitude is highest near the bed and decreases monotonically towards the free surface. Also, near the bed high-frequency fluctuations (corresponding to small-scale turbulence) can be seen to be superimposed on lower-frequency fluctuations. Near the surface, the high-frequency fluctuations are absent and the motion consists only of larger scales. In Hinterberger et al. (2008) spectra are provided which support these findings quantitatively. The lowest, thick-line signal in Figure 9.1 (right) represents the instantaneous depth-averaged u-fluctuations, showing clearly that also the depth-averaged velocity field carries fluctuations which a realistic 2D depth-averaged LES would have to yield. Vectors of the horizontal velocity fluctuations at the surface itself are displayed for one instant in Figure 9.2.

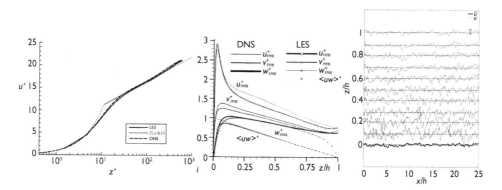

Figure 9.1 Flow in straight open channel with $Re_\tau = 590$. From Hinterberger et al. (2008), reproduced with kind permission from Springer Science + Business Media B.V.; DNS results of Moser et al. (1999). left: distribution of mean velocity and fluctuating components; right: streamwise variation of u-fluctuations at various depths for one instant in time.

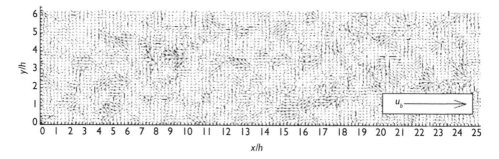

Figure 9.2 Flow in straight open channel with $Re_\tau = 590$: Vectors of velocity fluctuations at surface at one instant in time. From Hinterberger et al. (2008), reproduced with kind permission from Springer Science + Business Media B.V.

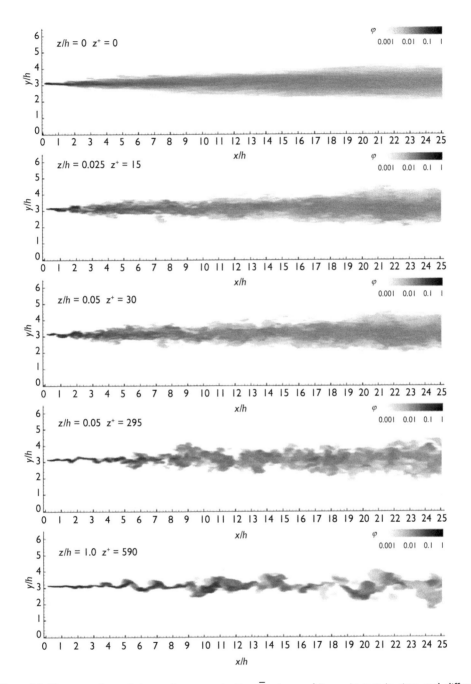

Figure 9.3 Contour plots of the scalar concentration $\bar{\phi}$ at an arbitrary instant in time and different depths indicated in each graph. The gray scale is logarithmic and ranges from 0.001 to 1. From Hinterberger et al. (2008), reproduced with kind permission from Springer Science + Business Media B.V.

They indicate the structures as well as their sizes and regions with relatively strong near-radial movement resulting from the impingement of eddies from below on the surface.

In addition to the 3D LES of the flow field, the development of a concentration field due to a tracer discharge from a vertical line source was also simulated. This field was determined by solving the transport equation (2.11) for the resolved scalar $\bar{\phi}$, with the subgrid-scale flux q_i^{SGS} calculated using an eddy-diffusivity model in which the SGS eddy diffusivity Γ_t is related to the SGS eddy viscosity v_t via $\Gamma_t = v_t/Sc_t$, employing a turbulent Schmidt number $Sc_t = 0.7$. A large molecular Schmidt number was used so that the molecular diffusion was always negligible compared with the SGS diffusion. The boundary conditions for solving the $\bar{\phi}$-equation were as follows. At $x = 0$ and in the middle of the channel ($y = \pi h$) the tracer was given in one horizontal computational cell and over the full depth the value $\bar{\phi} = 1$. Outside these middle cells the value of $\bar{\phi}$ was set to zero at the inflow boundary. At the outflow plane a convective boundary condition was used and at both bed and free-surface zero mass flux was prescribed. On the side boundaries in the transverse direction periodic conditions were used.

Figure 9.3 shows typical distributions of the calculated scalar concentration for one instant at different depths. The distribution of depth- and time-averaged concentration is given in Figure 9.4. This compares well with the corresponding experimental distribution reported by Rummel et al. (2005). The structures made visible by the concentration distributions in Figure 9.3 show clearly that the turbulent length scales vary significantly from the near-bed to the free-surface region. Near the bed the turbulence carries only small scales (leading to fairly uniform turbulent diffusion of the tracer) while near the surface the tracer is concentrated in larger structures. This confirms the findings obtained already from the velocity field. From the time-averaged transverse tracer spreading a transverse mixing coefficient Γ_y can be calculated (see Hinterberger et al. 2008), whose dimensionless value $\Gamma_y^* = \Gamma_y/u_* h$ has been measured in many dye-spreading experiments in flumes. Values in the range $\Gamma_y^* = 0.1\text{--}0.2$ are reported (Fischer et al. 1979), the large scatter resulting from the significant influence which small transverse mean motions may have. The LES reported here (Figure 9.4) yield a value of $\Gamma_y^* = 0.1$ which corresponds to the lower end of the

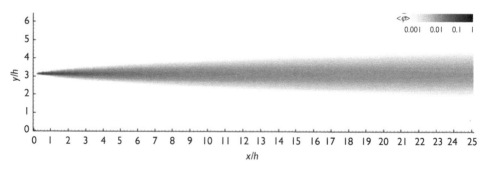

Figure 9.4 Depth-averaged and time-averaged concentration of the scalar $<\bar{\phi}>$. From Hinterberger et al. (2008), reproduced with kind permission from Springer Science + Business Media B.V.

experimental results and appears realistic as transverse mean motion was completely absent in the LES.

9.2 FLOW OVER ROUGH AND PERMEABLE BEDS

The study of turbulent flow over rough surfaces is highly important in hydraulic engineering and is an active area of research because practically all surfaces in open-channel flow must be considered rough. In addition to roughness, gravel-bed rivers feature surface permeability, and important mass and momentum exchange processes take place in a thin layer of the channel bed. Compared to flows over a smooth surface such flows are less well understood. Numerical studies resolving the individual roughness elements explicitly through the numerical grid can be very useful in providing detailed information of the flow close to the rough bed. Some DNS and LES for simple, well defined roughness geometries were discussed already in the Boundary Conditions chapter 6. Examples include the LES of Yang and Ferziger (1993) and Stoesser and Nikora (2008) of flow over 2D square bars mounted on a smooth wall. These and similar DNS studies (see chapter 6) provided a good picture of the near-wall flow and enhanced the understanding and quantification of turbulent flows over rough walls. However, an actual rough open-channel bed is represented more realistically by individual three-dimensional roughness elements closely packed together in a specific pattern. Resolving such roughness elements, Stoesser et al. (2003) performed a LES of open-channel flow over a bed roughened by a matrix of staggered cubes, and Singh et al. (2007) a DNS of the flow over a rough bed comprised of one layer of spheres. Alternatively to resolving the flow around simple roughness elements, the bed roughness can be accounted for by a suitable roughness model (see chapter 6 for such models), an approach which is computationally less demanding. In contrast to the numerous experimental research efforts to quantify the effects of roughness on the mean-flow statistics over impermeable rough beds/walls (see the summary by Jiménez, 2004), studies of turbulent flow over permeable beds are rather scarce and focused mainly on the determination of the vertical velocity profile above the permeable bed (e.g. Dancey et al., 2000); only Detert et al. (2010) also measured pressure and velocity signals within the permeable bed. Directly resolving the flow within the pores of a permeable bed is extremely complex and expensive, especially for natural permeable beds.

In the following, mainly two examples of LES in which the roughness is resolved explicitly through the grid will be presented, one for flow over a rough, impermeable bed and one with a permeable bed. A few results are added for LES with the roughness modelled. The two examples are taken from Bomminayuni and Stoesser (2011) and Stoesser et al. (2007). The computational set-ups of the LES, in terms of bulk Reynolds number and relative submergence of the roughness, were selected to be similar to the DNS of Singh et al. (2007) and laboratory experiments of flow over single and multiple layers (three and five) of spheres (Manes et al., 2009). The data of Singh et al. (2007) and Manes et al. (2009) were used to validate the statistical results of the LES. In the two LES discussed here, the channel beds were artificially roughened by closely packed hemispheres (impermeable bed) or spheres (permeable bed), respectively, with a diameter, D, and a flow depth above the hemisphere/sphere tops of $H \approx 3.4D$. The Reynolds number based on bulk velocity U(bulk) and channel depth H was Re $\approx$ 15,000 for both cases. The flows were driven by a pressure gradient dp/dx that unambiguously provided the global friction velocity, u_*, using $\tau_w = dp/dx \cdot H$. Based on this velocity ($u_* = \sqrt{\tau_w/\rho}$) and the flow depth (H), this corresponded to a friction Reynolds number of Re$_\tau \approx$ 1,200. Figure 9.5 presents the set-up of the permeable-bed simulation (**pbs**). The permeable bed consisted of three layers of spheres arranged in a cubical packing, placed on a smooth wall. The computational domain of the **pbs** spanned over $5.3H \times 3.5H \times (H + 3D)$ and included the surface flow region (H) and the subsurface region consisting of the 3 layers of spheres (Fig. 9.5). The computational domain of the impermeable bed simulation (**ibs**) spanned $6.12H \times 3.06H \times (H + 0.5D)$. The size of the domain was chosen large enough to include all relevant turbulence structures and was close to the often adopted $2\pi H \times \pi H \times H$ domain size for smooth-bed flows. High-resolution grids were employed consisting of 75 (**pbs**) and 91 (**ibs**) million grid points in total. The grid was Cartesian and each sphere/hemisphere was resolved explicitly using 40 grid points over the diameter in the horizontal and 60/30 grid points in the vertical direction (Fig. 9.6). Based on the friction velocity, the grid spacings in terms of wall units were $\Delta x^+ \approx 7$ in streamwise direction and $\Delta y^+ \approx 7$ in spanwise direction and $\Delta z^+ \approx 1.0 - 2.0$ close to the top of the spheres/hemispheres. Both flows were considered fully developed and in an infinitely wide domain, which allowed the application of periodic boundary conditions in the streamwise and spanwise directions. In both LES, the free surface was set as a frictionless rigid lid and was treated as a plane of symmetry, a reasonable approximation considering the low Froude number (Fr $\approx$ 0.17) of the flows. On both the sphere/hemisphere surfaces and the wall parts between the hemispheres, the no-slip condition was applied, and in the case of the spheres/hemispheres this was accomplished by employing the third-order accurate immersed boundary method of Peller et al. (2006). Both LES used the dynamic model due to Germano et al (1991) to compute the SGS-stresses.

The validation of the simulations was accomplished by comparing computed statistical quantities with measured data. Figure 9.7a presents streamwise velocity profiles of Bomminayuni and Stoesser's LES together with experimental data on a Clauser plot.

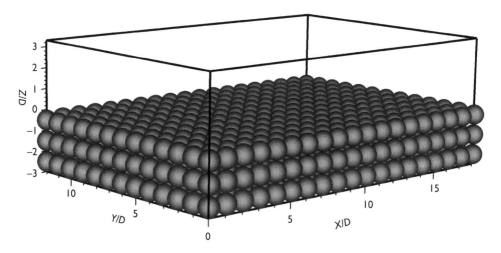

Figure 9.5 Flow over permeable bed consisting of 3 layers of spheres: Computational domain and arrangement of spheres. (from Stoesser et al., 2007)

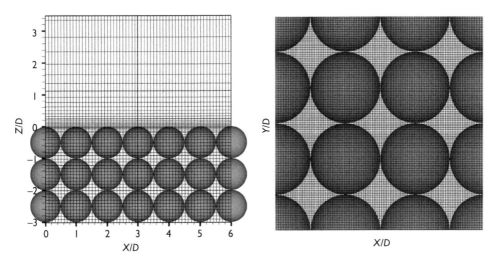

Figure 9.6 Flow over permeable bed consisting of 3 layers of spheres: Details of the grid in a vertical plane (left) and a horizontal plane cutting through the centres of the spheres (right). In the left plot only every 5th grid line is shown, while in the right plot every gridline is displayed. (from Stoesser et al., 2007)

Also plotted in this figure are the results of a LES of channel flow over a smooth bed, which illustrates the effect of roughness on the velocity distribution, i.e. a downshift of the profile. A log law was fitted to the rough-bed LES results (dashed line) and from this the downshift ΔB in Equation (6.18) was determined and then from ΔB the roughness Reynolds number $k_s^+ = k_s u_*/v = 95$ according to Equation (6.19). k_s is the equivalent sandgrain roughness. It can be seen that the LES results for $k_s^+ = 95$ agree quite well with the measured data for $k_s^+ \approx 100$. Figure 9.7b presents Clauser plots of velocity profiles for various k_s^+-values in the range 10 (transitional) to 257 (fully rough), again comparing LES results with experimental data. However, these LES did not resolve the individual roughness elements but were carried out by Stoesser (2010) with his roughness model introduced in Chapter 6 and served to validate this approach. The results demonstrate clearly the increasing downshift with increasing roughness height and show that the model can reproduce this in good agreement with experiments.

The vertical distribution of streamwise (u_{rms}) and wall-normal (w_{rms}) turbulent fluctuations (normalized by the global bed-shear velocity u_*) are presented in Figures 9.8a and 9.8b, respectively. The figures plot LES results (for both the rough and smooth channel flow) against DNS results of Singh et al. (2007) and various experimental data. The LES results are from the LES of Bomminayuni and Stoesser (2011) resolving the hemispherical roughness elements through the grid. The LES data are spatially averaged (in horizontal planes) results. It can be observed that the overall agreement between the LES results and the experimental/DNS ones is fairly good. The comparison of rough and smooth channel results indicates that the bed roughness influences the streamwise turbulence intensity over the entire channel depth. At regions close to the bed, this intensity is higher in the smooth channel but falls below as the distance from the bed increases. The peak in the spatially averaged turbulence intensity profile

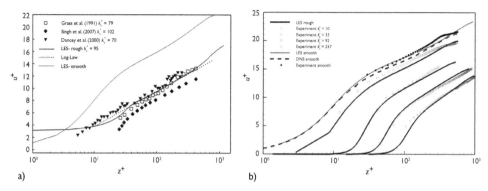

Figure 9.7 Flow over a bed roughened by hemispheres: Time and spatially averaged streamwise velocity profiles obtained from LES and data of previous experimental studies (Grass et al., 1991, Dancey et al., 2000) and DNS (Singh et al., 2007). Here, $u^+ = <u>/u_*$ and $z^+ = zu_*/v$. (a) from Bomminayuni and Stoesser (2011) in which the roughness elements were resolved explicitly, (b) from Stoesser (2010) in which the roughness was modelled, both reproduced with permission from ASCE.

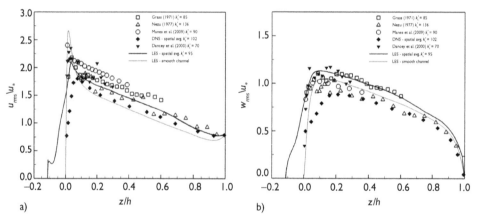

Figure 9.8 Flow over a bed roughened by hemispheres: Calculated and measured profiles of the (a) streamwise (b) wall-normal turbulence intensities of a roughness-resolved LES. Symbols are from experiments except for Singh et al. (2002) which is from DNS. The hemisphere top is located at $z/h = 0.028$. From Bomminayuni and Stoesser(2011), reproduced with permission from ASCE.

is located almost exactly at the roughness tops. The profiles of wall-normal turbulence intensity do not exhibit a significant peak. However, the maxima do not coincide with the peaks of streamwise turbulence intensities but are slightly shifted away from the wall. Unlike the streamwise turbulence intensity, the spatially averaged wall-normal intensity shows consistently higher values in the rough channel than in the smooth one. This implies that the roughness enhances mixing in the vicinity of the rough bed, with augmented turbulence further away from the wall. As a result, near-wall gradients of streamwise velocity remain high over a greater distance away from the wall.

Bed roughness reduces the peak value of streamwise turbulence intensity normalized by u_*, which can be shown by plotting this as a function of the roughness-height-to-water-depth-ratio as is done in Figure 9.9 for data from various experimental studies, direct numerical simulations (Scotti, 2006; Singh et al., 2007) and large-eddy simulations (Stoesser, 2010 and Stoesser and Bomminayuni, 2011). The data from both roughness-resolved-LES of Bomminayuni and Stoesser (2011, grey circular symbol) and the roughness-modelled-LES of Stoesser (2010, black circles) follow clearly the trend and are consistent with previous studies. The effect of greater roughness is enhanced mixing and momentum exchange near the rough wall with the consequence of spreading the turbulence over a wider area, thus reducing the peak turbulence intensity.

Bomminayuni and Stoesser (2011) educed turbulence structures over the rough impermeable bed by using contours of spanwise vorticity, velocity perturbation vectors and contours of the swirl strength in a longitudinal plane through the centre of hemispheres (Fig. 9.10). There are areas of strongly varying and intermittent vorticity throughout the depth (Fig. 9.10a), which suggest vortex structures in the flow. At regions close to the bed, the fact that the contour levels are predominantly positive indicates vertical shear (du/dz) but suggest also that these vortex structures rotate preferentially in the clockwise direction. Figure 9.10b shows a blow up of the flow near the rough bed, and in addition to vorticity contours velocity perturbation vectors are plotted, through which vortices can be discerned from vertical shear. Sweep (i.e. $u' > 0$ and $w' < 0$) and ejection (i.e. $u' < 0$ and $w' > 0$) events are present

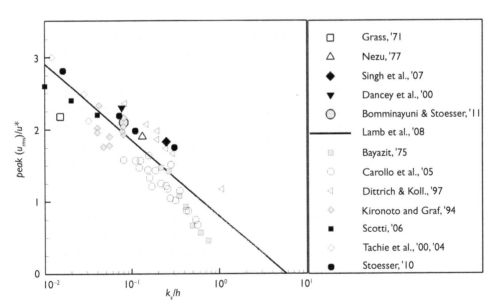

Figure 9.9 Flow over a bed roughened by hemispheres: Peak turbulence intensity as a function of the roughness-height-to-water-depth-ratio (from Bomminayuni and Stoesser, 2011, reproduced with permission from ASCE). The LES of Bomminayuni and Stoesser is a roughness-resolved simulation, while the LES of Stoesser (2010) uses a roughness closure model. Scotti (2006) and Singh et al. (2002) are DNS results while the other entries are experimental values. The black line is the empirical relationship proposed by Lamb et al. (2008).

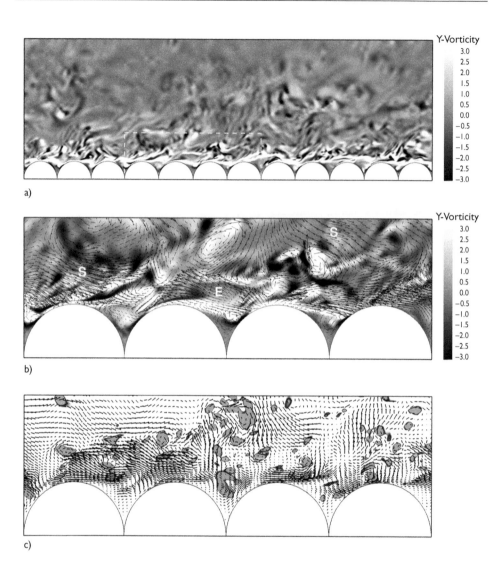

a)

b)

c)

Figure 9.10 Flow over a bed roughened by hemispheres: (a) Contours of instantaneous spanwise
vorticity in the flow in a longitudinal plane over the hemispheres (b) blow-up of a region
in which the vorticity contours are overlaid by vectors of the velocity fluctuation and (c)
swirl strength (grey areas) in a blow-up region close to the rough bed. From Bomminayuni
and Stoesser (2011), reproduced with permission from ASCE.

and are known to be the dominating turbulence structures that are mainly responsible for turbulent momentum transfer near the bed. For instance, in Figure 9.10b
such momentary events are marked with E (for ejection) and S (for sweep), during
which near-bed, low-momentum fluid is ejected into the boundary layer flow and
high-momentum outer-layer fluid is swept towards the roughness elements. Sweeps
and ejections are accompanied by hairpin vortices, signatures of which are also visible

in Figure 9.10b. A more quantitative way of identifying signatures of hairpin vortices is to plot the swirl strength, which is computed from a reduced velocity tensor that contains only the in-plane velocity gradients (Adrian et al., 2000a). Contours of the two-dimensional swirl strength isolate regions that swirl about the axis normal to the plane. Figure 9.10c presents contours of the swirl strength together with velocity perturbation vectors at an instant in time. A number of vortices, possibly hairpin packets, are identified through the swirl strength criterion and their distribution is similar to the one observed by Adrian (2007).

The existence of hairpin vortices (and others) is supported by the three-dimensional visualization of these vortical structures through an iso-surface of the 'Q-criterion' (Hunt et al., 1988) introduced in Chapter 8. In preliminary studies of the flow over a single hemisphere, the dominating coherent structures were educed by using the Q-criterion (see Fig. 8.12) and Horseshoe-Vortex (HSV) as well as Hairpin-vortex (HP) structures were identified.

Figure 9.11 presents a snapshot of an iso-surface of Q over a selected region of the hemispherical bed, in which structures similar to the ones just mentioned can be identified. In Figure 9.11, hairpin structures are marked as HP-1 and HP-2, and there are a number of elongated structures which are either remaining legs of hairpin vortices or streamwise vortices from which secondary hairpin vortices evolve. Also present are horseshoe-vortex-type structures (indicated as HSV-1 and HSV-2 in Fig. 9.11). These vortices are generated in a similar manner as the horseshoe-vortex system in Figure 8.12, except that here the vertical location of these HSVs coincides with the virtual bed location.

The turbulence structures over the permeable bed consisting of 3 layers of spheres are similar to the ones discussed above and are suggested to be the driving mechanism of mass and momentum exchange between the fast-moving outer flow and the very slow-moving flow in the upper layers of the permeable bed, often referred to as the hyporheic zone. Figure 9.12 provides snapshots of the complex velocity-perturbation field in two planes, one with maximum porosity (right) and one with minimum porosity (left), and gives evidence of the turbulence structures prevailing. At the given instant in time, ejections, transporting low-momentum fluid away from the permeable bed occurred, while at other instants sweeps, which convect high momentum fluid towards the bed, were observed. These events are found to be the main drivers for the hyporheic exchange as they force fluid out of the cavities (or pores) between spheres during ejections and promote inrushes of "fresh" fluid into the pores of the bed during sweeps. The latter can be seen quite well from the right part of Figure 9.12, which also depicts the turbulent motion in the pores induced by this exchange. This motion is however strongly damped with increasing distance from the bed surface.

The raw velocity time signals plotted in Figure 9.13 provide evidence of the highly fluctuating nature of the flow above the permeable bed (top left) but also in the pores, in particular in the first pore below the bed surface (top right). Even in the second pore turbulence is still visible but is already damped considerably compared to the flow outside the bed. The recurrence of strong sweeps and ejections is particularly obvious from the velocity signal outside the bed (top left) and sweeps and ejections are still visible in the first pore (top right). The damping of turbulence is quantified by the distribution of pressure fluctuations depicted in the bottom right part of Figure 9.13. This curve has a pronounced peak slightly below the top of the spheres, which is

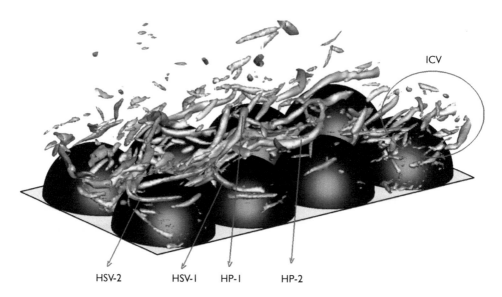

Figure 9.11 Flow over a bed roughened by hemispheres: Snapshot of an iso-surface of the Q-criterion over a selected region of the hemisphere bed showing different structures: Hairpin vortices (HP-1 and HP-2), Horse-Shoe Vortices (HSV-1 and HSV-2), Incoherent Vortices (ICV) are identified in the figure. From Bomminayuni and Stoesser (2011), reproduced with permission from ASCE.

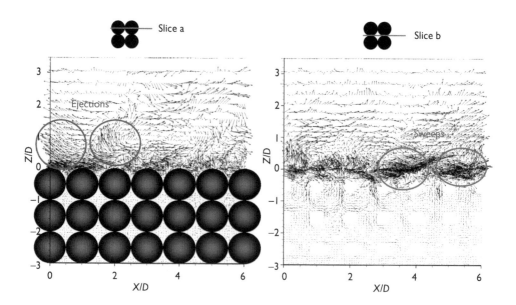

Figure 9.12 Flow over a permeable bed consisting of 3 layers of spheres: Perturbation velocity vectors ($u' - w'$) in two selected longitudinal planes: one with maximum (left) and one with minimum (right) porosity. (from Stoesser et al., 2007)

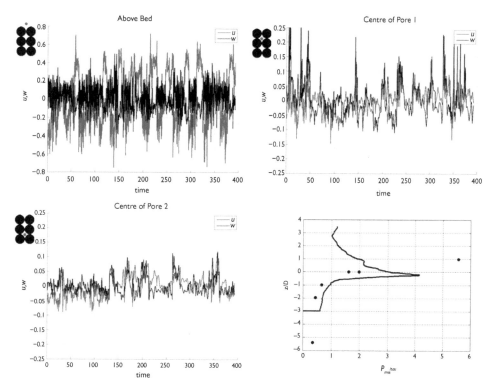

Figure 9.13 Flow over a bed consisting of 3 layers of spheres: Raw time series of streamwise (red) and wall-normal (black) velocity fluctuations at three locations: above the bed (top left), inside the first pore (top right) and inside the second pore (bottom left). Spatially averaged vertical distribution of pressure fluctuations (bottom right); Continuous line: LES, symbols: Experiment of Detert et al. 2010. (from Stoesser et al., 2007)

where the maximum pressure forces on the spheres are found. For comparison, experimental data of Detert et al. (2010) are included in the plot, which were obtained by measuring pressure signals in a naturally packed gravel bed. Simulated and measured decay of pressure fluctuations with the depth are very similar and follow an exponential curve. However, the level of pressure fluctuations in the experiment is somewhat lower due to the much denser packing of natural gravel in comparison to the artificial-sphere arrangement in the LES. The large experimental value at $z = D$ can be explained by the fact that this results from a point measurement at a single location (probably exactly above a gravel), while the LES curve represents a spatial average. However, values of pressure fluctuations up to 6 times larger than the mean wall shear stress were also observed at individual locations in the LES.

Although roughness-resolved large eddy simulations as presented here are extremely costly in terms of computational resources, the two examples have not only shown that LES returns good accuracy in terms of mean flow and turbulence statistics but also provides information on the highly fluctuating velocity and pressure field and thus valuable insight into physical mechanisms that are very difficult to deduce from experiments. This detailed information, including high-resolution spatial and temporal velocity and pressure distributions around each sphere, is extremely useful for studying the initiation of sediment transport and does in fact allow for a direct simulation of this process. Indeed, Chan-Braun et al. (2012) used DNS to compute directly forces and torque on individual particles comprising a rough bed and reported that particularly strong lift forces lead to critical Shields numbers. On the other hand, the computationally less demanding approach of modelling the bed roughness within LES offers the possibility to simulate accurately more complex rough-bed flows, e.g. flow around hydraulic structures in a rough-bed channel or the flow in an alluvial channel for which the exact details of the bed are unknown.

9.3 FLOW OVER BEDFORMS

The flow over bedforms in sand-bed rivers has been studied widely in order to understand the effect of bed shape on the hydrodynamics and turbulence and the mechanisms of bed-load and suspended-load transport. Already in the 1960's flume studies were carried out by Simon and Richardson (1963) and Guy et al. (1963), who categorized bed forms into ripples, dunes and antidunes. The importance of the hydrodynamic and geomorphological impact of bedforms in rivers is reflected in numerous experimental studies on dunes undertaken by Müller and Gyr (1986), Mierlo and Ruiter (1988), Lyn (1993), Bennett and Best (1995), Coleman and Melville (1996), Kadota and Nezu (2002), Hyun et al. (2003), Maddux et al. (2003a, 2003b) as well as many others. A wide range of configurations were investigated and progress has been made in understanding the flow and turbulence as well as the sediment transport characteristics (an extensive summary is given in a detailed review by Best, 2005). Because of bed deformation, the mean-flow characteristics, bed shear stresses, turbulence intensities and Reynolds Stresses as well as turbulent flow structures differ significantly from those over flat smooth or rough beds. For instance, dune formation causes the flow to separate at the dune crest, creating a large separation zone on the leeside of the dune. Associated with the separation zone is a turbulent free shear layer generating large scale eddies that travel through the flow domain and towards the surface while dissipating. The geomorphological origin of ripples and dunes is believed to lie in the presence of coherent structures, which are the driving mechanism for sediment transport and hence for the processes causing bed deformation. However, coherent structures are difficult to detect experimentally, as they vary strongly in space and time. Ideally, experimental investigations should be complemented with numerical simulations for the further exploration of the mechanisms of the instantaneous flow but also to serve as predictive tools to perform parametric studies. Until recently, only a few numerical simulations of the flow over dunes existed, which were mainly based on RANS methods (e.g. Mendoza and Shen, 1990, Yoon and Patel, 1996). RANS models can capture much of the mean-velocity information but fail to model the effect of the large-scale flow structures. Recent work applied LES to flow over dunes (Yue et al. 2005a and 2005b, 2006; Stoesser et al., 2008; Grigoriades et al., 2009, Omidyeganeh and Piomelli, 2011), which reproduced accurately turbulence statistics and revealed all relevant turbulence structures over dunes including rollers, kolk-boil vortices, splats and streaks. Chou and Fringer (2010) went one step further and coupled

LES with a sediment-transport and bedform-evolution model. They simulated the formation and evolution of sand ripples induced by turbulence in an oscillatory flow. Chou and Fringer (2010) demonstrated the importance of resolving and accurately computing the near-bed turbulence structures for the correct prediction of bed-form initiation and development.

In the following, some results of two recent LES of flow over two-dimensional dunes by Stoesser et al. (2008) and Omidyeganeh and Piomelli (2011) are presented. The set-up was identical for both LES and was selected to represent laboratory experiments undertaken by Polatel (2006). In these, a train of 22 two-dimensional fixed dunes was attached to the bottom of a flume. The dune height is $k = 20$ mm, and the dune wavelength is $\lambda = 400$ mm, so that $\lambda/k = 20$, which is in accordance with many previous studies and with dunes found in rivers. The two LES studies featured a dune-length-to-water-depth-ratio of $\lambda/h = 5$ (or water-depth-to-dune-height-ratios of $h/k = 4$, respectively). The Reynolds number Re, based on the average bulk velocity $U(bulk) \approx 0.3$ m/s and the maximum flow depth h, was approximately Re = 25,000. A two-component LDV system was used in the experiments to measure streamwise and vertical velocities at six selected verticals along the 17th dune. These data were used for comparing numerical simulations and experiments.

The computational domain of Stoesser et al. (2008) spanned one dune length λ in the streamwise, $4h$ in the spanwise and h in the vertical direction, respectively. Omidyeganeh and Piomelli's domain also spanned one dune length but was twice as that of Stoesser et al. in the spanwise direction. Stoesser et al.'s grid consisted of $416 \times 170 \times 128$ grid points in the streamwise, spanwise and vertical direction, respectively, with maximum grid spacing in terms of wall units of $\Delta x^+ \approx 20$ in streamwise direction, $\Delta y^+ \approx 19$ in spanwise direction and $\Delta z^+ \approx 1$ near the dune surface. Omidyeganeh and Piomelli used the same number of grid points in streamwise and wall normal direction and employed approximately two times the number grid points in the spanwise direction than Stoesser et al. because of their wider domain. Both LES resolved the near-wall flow with 3–6 points in the viscous sublayer, which justified the use of a no-slip wall condition. Also in both LES, the free surface was treated as a flat plane of symmetry (with zero stress condition) and periodic boundary conditions were applied in the streamwise and spanwise directions. Both LES used the dynamic Smagorinsky model to compute SGS stresses, and whilst Stoesser et al. (2008) used local averaging to obtain the Smagorinsky coefficient, Omidyeganeh and Piomelli (2011) made use of the Lagrangian averaging technique proposed by Meneveau et al. (1996).

Figure 9.14a presents streamlines visualizing the time-averaged flow and the size of the mean recirculation zone that forms behind the dune crest. Figure 9.14b compares measured and simulated time-averaged streamwise velocities along the six measurement verticals, which are drawn into Figure 9.14a. Omidyeganeh and Piomelli's data are represented by the solid line, while Stoesser et al's data are represented by the dashed line. The overall agreement is quite satisfying for both simulations and the two LES are virtually identical. In Figure 9.14c, the Reynolds shear stress is compared along the six verticals and again overall very good agreement was obtained with experiments.

Figure 9.15 shows turbulence structures at an instant in time, visualized by iso-surfaces of the instantaneous pressure fluctuation. Spanwise vortices (rollers) are generated in the shear layer separating from the crest due to a Kelvin–Helmholtz (K-H)

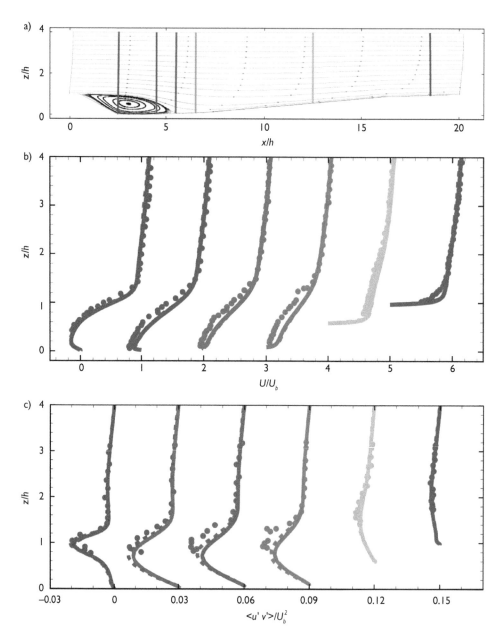

Figure 9.14 Flow over periodic dunes: (a) Streamlines of the mean flow and location of 6 measurement verticals at which statistical quantities are compared; measured (dots) and simulated time-averaged streamwise velocities (b) and Reynolds-shear-stresses (c) along the
six measurement verticals. The solid lines represent the LES results of Omidyeganeh
and Piomelli (2011), the dashed lines represent the LES results of Stoesser et al. (2008).
(from Omidyeganeh and Piomelli, 2011)

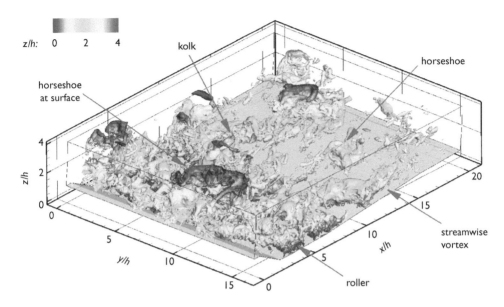

Figure 9.15 Flow over periodic dunes: Instantaneous flow structures at an arbitrary instant in time visualised with isosurfaces of pressure fluctuations. (from Omidyeganeh and Piomelli, 2011)

instability. They are convected downstream and either interact with the wall and then rise to the surface as so called kolk-boil vortices or are convected along the separated shear layer while diffusing. The kolk-boil vortices usually take the form of a horse-shoe or hairpin.

Figure 9.16 presents snapshots of the instantaneous flow visualized by contours of the vorticity magnitude (a) and velocity perturbation vectors (b) in a longitudinal plane at an instant in time. The rollers spring off the crest of the dune, and an area of high vorticity is visible downstream of the crest. The turbulence is strongest in the reattachment region, which is where most rollers impinge on the bed, generating kolk-boil vortices. The latter are convected towards the water surface where they eventually impinge as boils.

Instantaneous velocity vectors in a wall-parallel plane at approximately $\Delta z^+ \approx 10$ are presented in Figure 9.17a. Most turbulence occurs in the vicinity of the time-averaged location of reattachment ($X/k \approx 5$). The points of instantaneous reattachment are identified by areas of strongly diverging vectors (i.e. singular points) and of high pressure (Fig. 9.17b), where impinging fluid is reoriented in all directions tangential to the wall. These splats are a result of the impermeability condition of the wall, with the consequence of the wall-normal component of turbulent kinetic energy being transferred to the tangential components. In this region, the spanwise fluctuations were found to be of the order of the streamwise fluctuations or even higher (Stoesser et al., 2008). Further downstream of the mean reattachment region, the footprints of vortices convected downstream are revealed by smaller splats as well as areas of streaks so that areas of high and low instantaneous pressure exist (Fig. 9.17c). Further downstream these vortices are lifted upwards as the boundary

a)

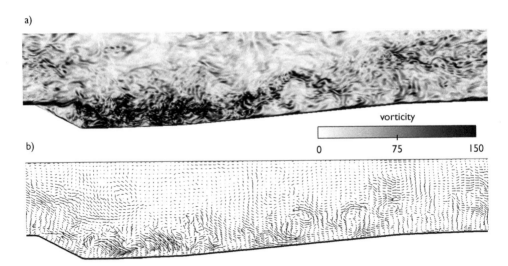

b)

Figure 9.16 Flow over periodic dunes: (a) Distribution of the magnitude of the vorticity vector at an instant in time and (b) velocity perturbation vectors in a selected longitudinal plane. From Stoesser et al. (2008), reproduced with permission from ASCE.

layer grows and the splats disappear while the near-wall streaks remain. This can also be seen clearly in Figure 9.18 where instantaneous streamwise velocity fluctuations in a wall-parallel plane near the bed are visualized. Again, in the area of the mean reattachment strong fluctuations without a clear structure can be seen, and further downstream from about $X/k \approx 10$ elongated streamwise structures representing high- and low-speed streaks form and are well known features of near-wall attached flow in boundary layers and channels.

A horseshoe-shaped kolk-boil vortex as it rises to the surface is visualized with an isosurface of the pressure fluctuation in Figure 9.19. Eventually the vortex reaches the water surface (lower part of Fig. 9.19) and creates an upwelling there, a so-called boil. The vortex is originally vertical, but is later tilted at an angle of 40–60 degrees with respect to the vertical axis. Similar visualizations of and findings on the kolk-boil vortex system were reported in Grigoriades et al. (2009), who performed a large-eddy simulation of the flow over a 2D dune similar to the one reported on here.

A boil scenario is quantified with the help of Figure 9.20, in which contours of the pressure-fluctuation p' and velocity perturbation vectors at the water surface at an instant in time are plotted (right part of Fig. 9.20). A region of strong pressure fluctuation p' is observed, which is the result of the head of the horseshoe vortex impinging on the water surface (circle in Fig. 9.20). As a result of impingement, the perturbation vectors are oriented radially away from the point of impingement.

The LES of flow over a two-dimensional dune of Stoesser et al. (2008), Omidye-ganeh and Piomelli (2011) and also the ones reported in Grigoriades et al. (2009) and Yue et al. (2005, 2006) revealed the prevailing turbulence structures and have contributed significantly to the understanding of the generation and fate of large-scale coherent structures in flows over bedforms.

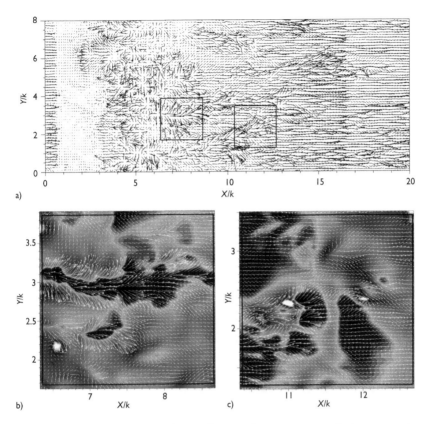

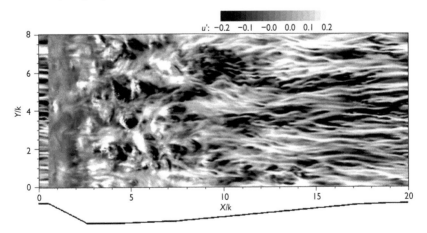

Figure 9.17 Flow over periodic dunes: (a) Distribution of instantaneous velocity vectors in a wall-parallel plane near the bed and (b) and (c) blow-ups at selected locations showing velocity vectors and contours of high (dark areas) and low (light areas) pressure. From Stoesser et al. (2008), reproduced with permission from ASCE.

Figure 9.18 Flow over periodic dunes: Distribution of instantaneous streamwise velocity fluctuations in a wall-parallel plane near the bed at $\Delta z^+ \approx 10$. From Stoesser et al. (2008), reproduced with permission from ASCE.

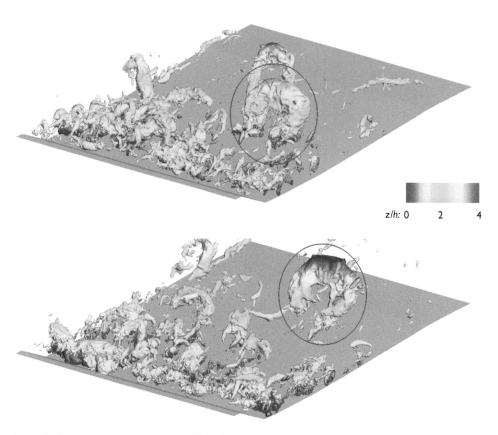

z/h: 0　　2　　4

Figure 9.19 Flow over periodic dunes: Turbulence structures at two instants in time visualised with isosurfaces of pressure fluctuation. (from Omidyeganeh and Piomelli, 2011)

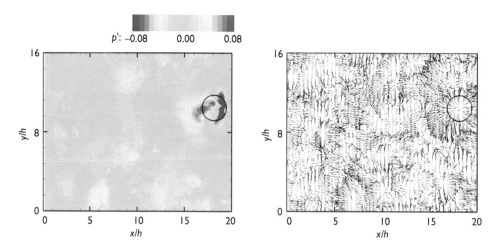

Figure 9.20 Flow over periodic dunes: Contours of instantaneous pressure fluctuation (left) and velocity perturbation vectors (right) at the water surface. (from Omidyeganeh and Piomelli, 2011)

9.4 FLOW THROUGH VEGETATION

The presence of vegetation on banks and floodplains of rivers and streams influences considerably the horizontal and vertical distributions of mean and fluctuating velocity, and hence also of turbulence quantities as well as the transport of sediments and solutes. The flow through partially-vegetated channels or emergent and submerged vegetation is characterized by significant velocity gradients (laterally, longitudinally and vertically) resulting in shear-layer formation and large-scale turbulence-structure generation. A large number of experimental studies were undertaken with the goal of quantifying flow resistance in terms of a drag force (e.g. Stone and Shen, 2002; Tanino and Nepf, 2008), which was found to be a function of vegetation density, vegetation geometry and stem Reynolds number. A number of experimental works have focused on detailed examination of the flow field and turbulence structure within a plant canopy, for which the vegetation was modeled as an array of rigid cylinders at regular spacing (Pasche and Rouve, 1985; Nepf and Vivoni, 2000; Lopez and Garcia, 2001; Ghisalberti and Nepf 2002; White and Nepf, 2008; Liu et al, 2008). Most of the numerical work on flow through vegetation to date has been based on the Reynolds-Averaged-Navier-Stokes (RANS) equations, in which the effect of vegetation on the mean flow and turbulence is estimated by adding sink/source terms to the momentum and turbulence transport equations to account for vegetative drag and turbulence-production effects. RANS models have been developed and validated by Shimizu and Tsujimoto (1994), Naot et al. (1996), Neary (2000), Fischer-Antze et al. (2001) and Choi and Kang (2001) among others. Even though the time-averaged flow was predicted to a very satisfying degree using RANS models, these were less successful in predicting turbulence-related quantities due to the fact that RANS models cannot accurately account for the organized large-scale unsteadiness and asymmetries (coherent structures) induced by the local flow-vegetation interaction. Recently, Large-Eddy Simulations (LES) of channel flow through vegetation were presented by Cui and Neary (2002, 2008) and Stoesser et al. (2009, 2010a, and Kim and Stoesser, 2011), which demonstrated that LES can elucidate the large-scale coherent structures and provide superior accuracy in the prediction of turbulence statistics.

Here an example of the utilization of LES to predict flow through emergent vegetation idealized by cylinders is presented. Stoesser et al. (2010a) resolved the flow around the individual cylinders by a high-resolution grid through which each vegetation element was explicitly represented, so that pressure and friction drag could be computed directly.

In Stoesser et al.'s (2010a) LES, calculated time-averaged velocity and turbulence-intensity profiles were first compared with laboratory data to validate the approach and chosen grid resolution. Afterwards, further LES were performed to investigate the effects of vegetation density and cylinder Reynolds number on flow resistance as well as on the time-averaged and instantaneous flow fields. The simulation set-up of the validation case was chosen to match the experiments carried out by Liu et al. (2008) who placed a matrix of rigid circular cylinders of diameter D in a staggered arrangement (see Fig. 9.21) into a rectangular flume and carried out detailed LDA measurements at the six verticals within the flow indicated in Figure 9.21a. In Figure 9.21, s is defined as the distance between two cylinders in the streamwise direction and in Liu et al.'s experiment $s = 10D$. The vegetation density φ, here defined as the volume occupied by

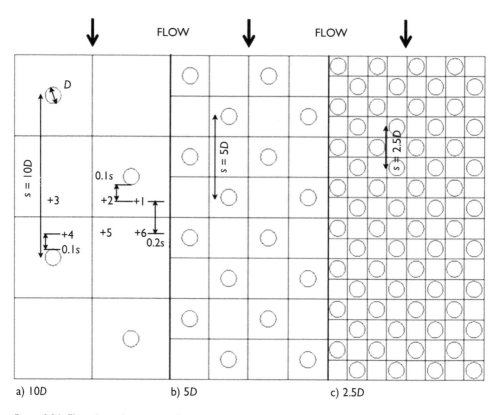

a) 10D   b) 5D   c) 2.5D

Figure 9.21 Flow through a matrix of emergent cylinders: Flow domain and cylinder arrangement for the three different computational set-ups. The six measurement locations of the experiment are depicted in (a). From Stoesser et al. (2010a), reproduced with permission from ASCE.

the cylinders divided by the total volume is $\varphi = 0.016$. The cylinder Reynolds number based on the bulk velocity, u_{bulk} and the cylinder diameter D is $\text{Re}_D = 1340$. In addition to the 10D case of Liu et al. (2008), LES were carried out at two additional vegetation densities i.e. $s = 5D$ and $s = 2.5D$, or $\varphi = 0.063$ and $\varphi = 0.251$, respectively (Fig. 9.21) and at one additional (lower) stem Reynolds number of $\text{Re}_D = 500$. In total, six different numerical experiments were performed. The computational flow domain chosen was the same for all cases and spanned 20D in streamwise, 10D in spanwise and 10.22D (corresponding to the water depth) in the vertical direction, respectively. A block-structured grid that was composed of several Cartesian H-grid and curvilinear O-grid blocks was employed, and the details of each grid for the six numerical experiments are summarized in Table 9.1. Stoesser et al. (2010a) applied the no-slip condition at the bed and at the cylinder surfaces which was justified by the fact that the first grid point off the wall was situated well within the viscous sublayer. Periodic boundary conditions were applied in the streamwise and spanwise directions, and the free surface was treated as a frictionless rigid lid at which a zero-stress boundary condition was used. The simulations were initially run for about 20 eddy turn-over time units, t_e, defined as the water depth over the friction velocity, in order to establish fully

Table 9.1 Flow through vegetation: Parameters of the numerical grids for the six cases investigated.

Re_D	Spacing s	Gridpoints in H-grid $n_x \times n_y \times n_z \times n_{set}$	Gridpoints in O-grid $n_\theta \times n_r \times n_z \times n_{set}$	Total number of gridpoints
1340	10D	$82 \times 82 \times 122 \times 4$	$(82 \times 4) \times 52 \times 122 \times 4$	11,604,640
	5D	$62 \times 62 \times 122 \times 16$	$(62 \times 4) \times 32 \times 122 \times 16$	22,994,560
	2.5D	$42 \times 42 \times 122 \times 64$	$(42 \times 4) \times 12 \times 122 \times 64$	29,514,240
500	10D	$62 \times 62 \times 122 \times 4$	$(62 \times 4) \times 47 \times 122 \times 4$	7,564,000
	5D	$42 \times 42 \times 122 \times 16$	$(42 \times 4) \times 27 \times 122 \times 16$	12,297,600
	2.5D	$22 \times 22 \times 122 \times 64$	$(22 \times 4) \times 9 \times 122 \times 64$	9,963,008

developed flow conditions. For the calculation of mean-flow and turbulence statistics the simulations were then continued for 50+ eddy turn-over time units.

In the validation phase of the study, i.e. for the 10D case, Stoesser et al. (2010a) compared time-averaged streamwise and vertical velocities and streamwise turbulence intensity profiles calculated by the LES with the ones measured in the experiment at six locations. It was found that the velocity profiles and the turbulence intensity profiles matched the Liu et al. (2008) observations quite well. Noteworthy is that Stoesser et al. (2010a) were able to reproduce accurately the near-bed bulge in the streamwise velocity profile, which is the result of the prevailing secondary flow that entrains high-momentum fluid into the wake near the bed.

Figure 9.22 presents contours of the time-averaged streamwise velocity and streamlines at about half depth for the 3 different densities investigated. In the 10D and 5D cases, there is a pronounced low-momentum wake behind each cylinder, and the wakes are clearly separated in the spanwise direction by a high-velocity corridor. The 2.5D flow field does not feature such separation, the wakes of upstream cylinders interfere and large velocity gradients in both streamwise and spanwise directions occur. The streamlines reveal that the 10D and 5D cases exhibit similar flow features, i.e. flow separation from the cylinder at approximately 90° and a relatively large recirculation region that is comprised of two counter-rotating vortices having about the length of the cylinder diameter. In the 2.5D case, the flow separates considerably later (at approx 130°) because the boundary layer on the cylinder is energized by the wake turbulence of upstream cylinders, and in this case the recirculation region behind the cylinder is much smaller than the ones found behind the cylinders of the 10D and 5D cases. The length of the vortex pair behind the 2.5D case cylinder is approximately only 1/4th of the cylinder diameter.

The explicit resolution of each stem through a very fine grid allowed a direct and accurate calculation of pressure and shear forces acting on the stems, so that the total drag force can be quantified. Stoesser et al. (2010a) compared the computed drag force on the cylinder with the drag forces of flow through emergent vegetation measured by Tanino and Nepf (2008), who carried out laboratory experiments to investigate the effect of cylinder Reynolds number and vegetation density on drag force and drag coefficient. The drag forces calculated by LES are plotted as a function of cylinder Reynolds number together with Tanino and Nepf's measured values in Figure 9.23. The LES complement effectively the previous experimental observations

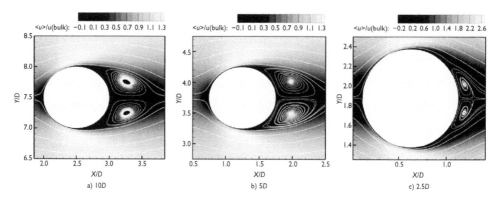

a) 10D   b) 5D   c) 2.5D

Figure 9.22 Flow through a matrix of emergent cylinders: Contours of time-averaged streamwise velocity in a horizontal plane at $Z/D = 0.5$ together with streamlines around one of the cylinders for the 3 cases at $Re_D = 1340$. From Stoesser et al. (2010a), reproduced with permission from ASCE.

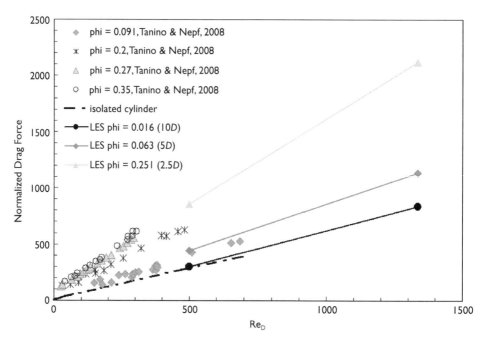

Figure 9.23 Flow through a matrix of emergent cylinders: Computed and measured normalized drag force as a function of cylinder Reynolds number for various vegetation densities. From Stoesser et al. (2010a), reproduced with permission from ASCE.

and match the observed trends very well. The drag-force distribution for the 10D case exhibits the same behaviour as an isolated cylinder (dashed line) suggesting that the flow recovers sufficiently behind each cylinder and that there is no wake interference. With increasing vegetation density and decreasing cylinder Reynolds number the normalized drag force increases.

The benefit of employing LES for such complex flows is not only that it reproduces reliably and accurately the turbulence statistics and the acting forces but als allows to study the instantaneous flow, thereby revealing important physical mechanisms and relationships. The horizontal distribution of vertical vorticity (z-vorticity) at an instant in time in a plane approximately at half water depth ($Z = 5D$) is presented for two vegetation densities in Figure 9.24. The flow through less dense vegetation (case $s = 5D$ on the LHS of Fig. 9.24) is characterized by regular vortex shedding. High levels of vertical vorticity are found close to the cylinders, but these are decreasing while the vortices are convected downstream. As the density of vegetation increases, the vortex shedding at the cylinders is influenced by upstream vortices resulting in a more irregular shedding behaviour. This is supported by elevated levels of vertical vorticity which are now also found on the upstream side of the cylinders in the 2.5D case (RHS of Fig. 9.24).

An impression of the 3D large-scale vortical structures behind the cylinders can be obtained from instantaneous iso-surfaces of the pressure perturbation $p' = (p - <p>)/(\rho U_\infty)$. Snapshots showing such iso-surfaces in an oblique view for the cases 10D and 2.5D are presented in Figure 9.25. For the 10D case (left), vertical vortices are observed on the sides of the cylinder originating from the separated shear layer behind the cylinder. The vortices extend over the full cylinder height and exhibit similar features as von Karman vortices behind long isolated cylinders, where the shear layers roll up and trigger regular alternating shedding of vortices. In the 2.5D case, the process is influenced strongly by the flow acceleration between the cylinders and the presence of vortices that are shed at upstream cylinders, which alters the shedding process and the structure of the vortices. As a result, the vortices do not extend over the entire water depth and appear to be incoherent.

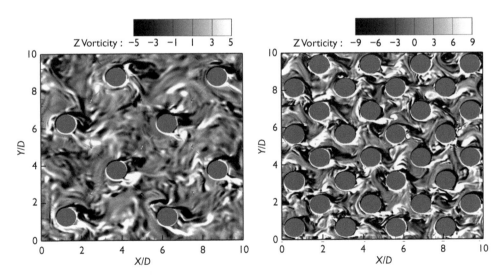

Figure 9.24 Flow through a matrix of emergent cylinders: Snapshot of contours of vertical vorticity in a horizontal plane at half channel depth for two vegetation densities. Left: 5D, Right: 2.5D. (Note that the 2.5D case has a different scale). From Stoesser et al. (2010a), reproduced with permission from ASCE.

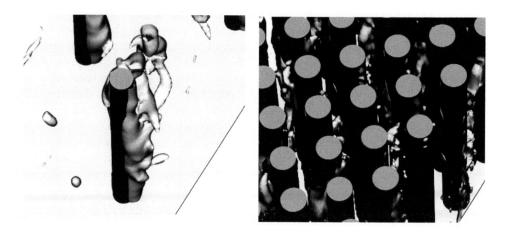

Figure 9.25 Flow through a matrix of emergent cylinders: Isosurfaces of pressure fluctuations for two vegetation densities (Left: 10D, Right: 2.5D). From Stoesser et al. (2010a), reproduced with permission from ASCE.

For the 10D case, snapshots of three-dimensional structures visualized with iso-surfaces of the Q-criterion (Hunt et al, 1998) from above (left) and by an oblique view (right) are presented in Figure 9.26. The above-mentioned regular shedding of von Karman vortices is clearly visible (indicated by lines connecting A-B-C, left). While these structures exhibit some two-dimensionality in their early stage, they are stretched in the streamwise direction as they are convected downstream. Packets of smaller vortices evolve, but these lose their coherence and weaken before reaching the downstream cylinder (Fig. 9.26 right).

Stoesser et al. (2010a) quantified the observed vortex-shedding behaviour in terms of a Strouhal number, St, and compared it with previous investigations. The Strouhal numbers of the six simulations performed in Stoesser et al.'s (2010a) LES study are plotted in Figure 9.27 together with data from experimental studies. The 10D case exhibits flow features that are very similar to the flow around an isolated cylinder, which is confirmed by the comparison of the St number as a function of Re_D. The values obtained herein match rather well the isolated cylinder values of Zhang and Dalton (1997), Kevlahan (2007) or Liu and Fu (2003) or the theoretical curve provided by Norberg (2003). As the stem density increases, there is an increase in Strouhal number but again no obvious dependency on the cylinder Reynolds number, which is confirmed by also plotting the experimental values of Lam and Lo (1992). However, for the 2.5D case the distribution of the Strouhal number as a function of Re_D suggests that there is a dependency of St on the cylinder Reynolds number at higher vegetation densities as the Strouhal number decreases with increasing Re_D. This is again in agreement with the observations of Lam and Lo (1992) for their 2.5D experiment.

In conclusion, the LES of Stoesser et al. (2010a) of flow through emergent vegetation presented here demonstrate the potential of LES for providing accurate turbulence statistics as well as for allowing the elucidation of turbulence structures and for shedding light on the important physical mechanisms at play.

Figure 9.26 Flow through a matrix of emergent cylinders: Instantaneous isosurfaces (coloured by the streamwise velocity) of the Q-criterion for the 10D case. Left: Top view, Right: Oblique view. From Stoesser et al. (2010a), reproduced with permission from ASCE.

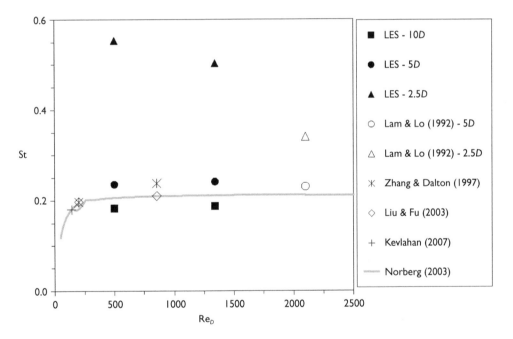

Figure 9.27 Flow through a matrix of emergent cylinders: Computed and measured Strouhal number as a function of cylinder Reynolds number versus isolated-cylinder and tube-bundle flow. From Stoesser et al. (2010a), reproduced with permission from ASCE.

9.5 FLOW IN COMPOUND CHANNELS

Compound channels are ubiquitous in the natural and man-made environment and consist of a main channel that carries the bank-full, more frequent discharges and a floodplain that is inundated during flood flows. The detailed distribution of flow and turbulence properties in such channels has recently found renewed attention due to an increase in flood extent and frequency and the concomitant flood damages. In compound open channel flow, mass and momentum transfer occurs at the interface between deep and shallow parts of the channel due to the presence of a mixing layer generated by the prevailing spanwise gradient of the primary velocity. The main drivers of the exchange are large-scale horizontal (quasi-2D) turbulence structures in the mixing layer (see upper part of Fig. 9.32), which are constantly fed by the lateral shear of the primary flow.

Most numerical simulations of flow in compound channels to date were based on the RANS equations (e.g. Naot et al., 1993, Pezzinga, 1994, Sofialidis and Prinos, 1998), and only very few studies employed LES (e.g. Thomas and Williams, 1995; Cater and Williams, 2008, Kara et al., 2012a). Whilst RANS methods with algebraic-stress or nonlinear eddy-viscosity turbulence models are able to predict quite well the time-averaged quantities, LES reproduces all unsteady flow features and thereby not only ensures great accuracy of turbulence statistics but also offers the possibility to elucidate dominating turbulence structures and their relevance for the statistics.

In the following, two examples of LES of flow and mass transport in compound channels are presented. The first example is a study by Kara et al. (2012a), who investigated the effect of floodplain depth on the flow and turbulence statistics in a compound channel. In the second example by Kara et al (2012b), LES was employed to investigate the effect of floodplain roughness on mass and momentum transfer in compound channels with shallow floodplains.

The computational set-up of the LES of Kara et al (2012a) was chosen to correspond to the experiments carried out in an asymmetric compound channel by Tominaga and Nezu (1991), whose data were used to validate the LES method employed. Two different floodplain depths were considered, i.e. $h/H = 0.5$ and $h/H = 0.25$, where h is the floodplain depth and H the main channel depth. The compound channel was quite narrow and main channel width, W, and floodplain width, w, were $W/H = 2.5$ and $w/H = 2.5$, respectively. The computational domain covered the full width and was $12H$ long, which ensured that even the largest turbulence structures could be captured. The Reynolds number, based on hydraulic radius and bulk velocity, was approximately $Re = 5 \times 10^4$. The flow was statistically homogenous in the streamwise direction and cyclic boundary conditions were employed in this direction. The compound channels had a smooth bed and smooth sidewalls and the no-slip wall boundary condition was used. The free surface was set as a frictionless rigid lid and was treated with a zero-gradient slip condition. The computational domain was discretized with a fine uniform grid of $601 \times 161 \times 251$ grid points for the main channel and 601×81 (41 for $h/H = 0.25$) $\times 251$ grid points for the floodplain in the streamwise (x), wall normal (y) and spanwise (z) directions, respectively. The grid spacings in wall units in the streamwise, wall-normal and spanwise directions were $\Delta x^+ \approx 26$, $\Delta y^+ \approx 6$ and $\Delta z^+ \approx 13$, in which u_* was determined from the pressure gradient dp/dx that drove the flow, which unambiguously provided the squared global friction velocity u_*^2.

Figure 9.28 presents velocity vectors of the secondary flow from the LES and from the experiments of Tominaga and Nezu (1991), for both floodplain-depth cases. In straight open-channels a secondary flow is generated as the result of turbulence anisotropy, and because LES simulates the energetic large-scale turbulence explicitly, the turbulence-driven secondary flow can be computed fairly accurately with LES. In the cases

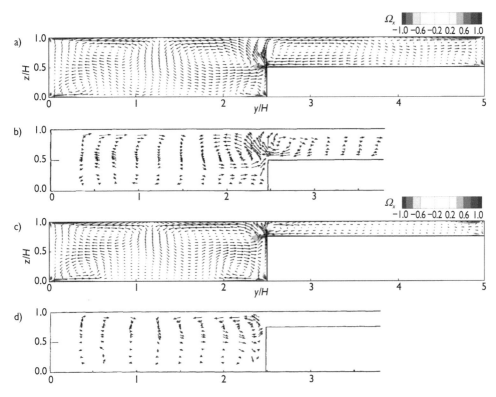

Figure 9.28 Flow in asymmetric compound channels: Streamwise vorticity and secondary currents from (a) LES, (b) experiment for $h/H = 0.5$ (c) LES, (d) experiment for $h/H = 0.25$. From Kara et al (2012a); Experimental results adapted from Tominaga and Nezu (1991).

investigated by Kara et al (2012a), a secondary-flow vortex pair forms at the interface between main channel and floodplain, regardless of the floodplain depth. Overall, the agreement of the predicted vortex pair with the one measured in the experiment is quite satisfying for both cases. It was found that in this particular compound-channel geometry, the fairly strong secondary current at the interface results in upwelling of fluid to the water surface, thereby disrupting any large-scale horizontal shear layer structures. In addition to the secondary-flow vectors, Kara et al. (2012a) also presented contours of streamwise vorticity, which are included in Figure 9.28. The color-coding is chosen such that only areas of significant positive (red) and negative (blue) streamwise vorticity are highlighted. The overall distribution is similar for the two flows, and vorticity magnitudes are greatest in the channel corners and in the vicinity of the interface, which is where the secondary currents originate (Kara et al., 2012a). There are areas of increased vorticity near the channel walls and the water surface, which is where the velocity normal to these boundaries is reduced to zero over a short distance.

Figure 9.29 presents simulated and measured isotachs of the time-averaged primary velocity, U, normalized with the maximum streamwise velocity, U_{max} for both floodplain depths. The LES predictions are in good agreement with the experiments, especially in

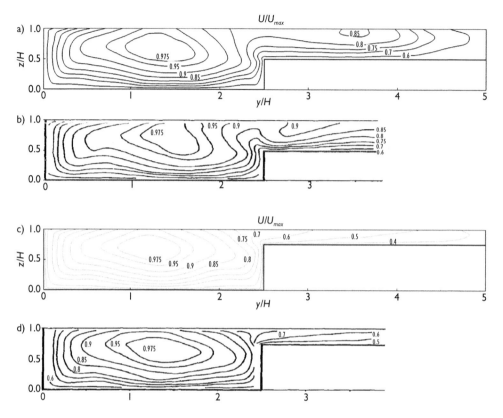

Figure 9.29 Flow in asymmetric compound channels: Contours of the primary velocity from (a) LES, (b) experiment for $h/H = 0.5$ (c) LES, (d) experiment for $h/H = 0.25$. From Kara et al (2012a); Experimental results adapted from Tominaga and Nezu (1991).

the main channel. Similar LES results and comparisons were reported in the studies of Thomas and Williams (1995) and Cater and Williams (2008), both of which focused on the deep floodplain case. The LES slightly underestimates the velocities on the floodplain, which could be due to insufficient flow development in the experiment. There is a substantial velocity gradient between main channel and floodplain flows, and the gradient is steeper in the shallow floodplain case than in the deep one. The primary velocity is influenced strongly by the prevailing secondary currents, most visible in the contour lines in the vicinity of the interface where they bulge at an inclined angle from the floodplain corner towards the free surface in the main channel.

Figure 9.30 presents the contours of the turbulent kinetic energy, k, normalized by the squared global friction velocity, u_*^2, of the LES of Kara et al. (2012a) and of the experiment for $h/H = 0.5$. The experimental data are available only for this case, and the LES captures the distribution of turbulent kinetic energy over the cross-section very well; however the LES values are generally slightly lower than the measurements, which is probably due to the fact that only the resolved turbulent kinetic energy is plotted. It is seen from the figures that the turbulent kinetic energy increases significantly

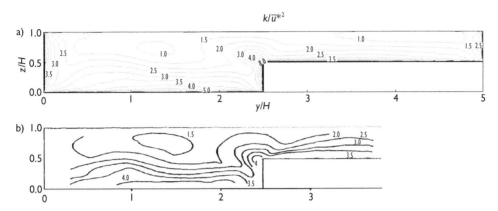

Figure 9.30 Flow in asymmetric compound channels: Distribution of normalized turbulent kinetic energy from (a) LES, (b) experiment for $h/H = 0.5$. Simulations unpublished, experimental results adapted from Tominaga and Nezu (1991).

as the interface is approached. In this region, the secondary flow at the juncture causes a bulging of the contours of k towards the free surface, indicating that turbulence is transported by means of the secondary flow.

Computational setup and boundary conditions of the LES of Kara et al (2012b) were identical to the ones reported in Kara et al. (2012a), however Kara et al. (2012b) focused on the shallow floodplain case, i.e. $h/H = 0.25$, and varied floodplain roughness to study its effect on turbulence quantities and mass and momentum transfer between deep and shallow parts of the channel. In addition to the smooth floodplain flow reported in Kara et al (2012a), referred to as case 1, Kara et al. (2012b) simulated floodplain flow over a rough bed, with a grain roughness height of $k_s/H = 0.04$ (case 4) and through emergent vegetation at two different vegetation densities (Cases 2 and 3) with solid volume fractions of $\phi = 0.08$ (Case 2) and $\phi = 0.04$ (Case 3), respectively. For the vegetation cases (2 and 3) the number of grid points was doubled in the streamwise direction over that used in Kara et al. (2012a), i.e. $\Delta x^+ = 13$, to enable proper resolution of the square vegetation elements, the wall treatment of which were accomplished by Werner and Wengle's wall model described in section 6.5. In addition to the flow, Kara et al. (2012b) simulated the transport of a tracer to investigate the effect of floodplain roughness on mass exchange between main channel and floodplain. The tracer was constantly fed to the flow over the entire floodplain width and depth at $x/H = 0.0$ (i.e. Dirichlet boundary condition) and the tracer exited the domain at $x/H = 16$ where a zero-gradient Neumann condition was used.

Figure 9.31 presents contours of the turbulent kinetic energy (tke) normalized with u_*^2 in the cross-section for two of the four cases investigated. Also plotted are vectors of the time-averaged secondary flow, illustrating the strong linkage between secondary flow and turbulence. Floodplain vegetation (case 3) substantially retards the flow on the floodplain, causing large spanwise gradients of streamwise velocity. This increases the lateral shear between main channel and floodplain, producing significant turbulence at the main-channel-floodplain-interface. In addition to the

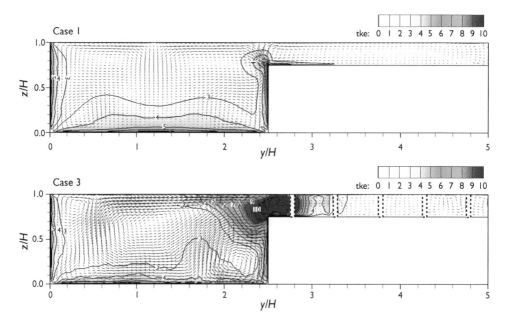

Figure 9.31 Flow in asymmetric compound channels: Cross-sectional distribution of longitudinally–averaged turbulent kinetic energy, tke, normalized by the squared friction velocity, u_*^2, for two different floodplain roughnesses (a) Case 1, smooth (b) Case 3, vegetated with phi = 0.04. From Kara et al (2012b).

interface shear, von Karman-type vortices are shed irregularly from the cylinders that are positioned closest to the interface (see bottom left of Fig. 9.32) enhancing the turbulence there. Hence, substantially elevated levels of the normalized tke over the entire water depth are observed in the vicinity of the interface. For a comparison, high values of tke are concentrated only around the corner of main channel and floodplain in the smooth floodplain case 1.

The upper part of Figure 9.32, which shows contours of the instantaneous velocity magnitude, provides a three-dimensional picture of the flow of case 3. The flow is characterized by a diversity of instantaneous flow structures. Firstly, large-scale horizontal free-surface structures prevail in the vicinity of the interface. These structures form as a result of the strong shear between the slow floodplain flow and the faster main-channel flow, promoting horizontal vortices and constantly feeding into them. The size of the $2D$ vortices depends on the magnitude of the lateral shear, i.e. the greater the shear the larger are the vortices (see Fig. 9.33 for visualization). Their growth is restricted by the bed shear stress and the corresponding three-dimensional turbulence and by the secondary flow near the interface, disrupting and diffusing these structures. Secondly, von Karman-type vortices are shed from each vegetation element. The oncoming flow separates periodically and in an alternating manner at the cylinder and vortices roll up in the so created shear layers and are then convected downstream with the flow. Close to the interface, this vortex shedding process is influenced by the aforementioned horizontal 2D structures and irregular shedding is

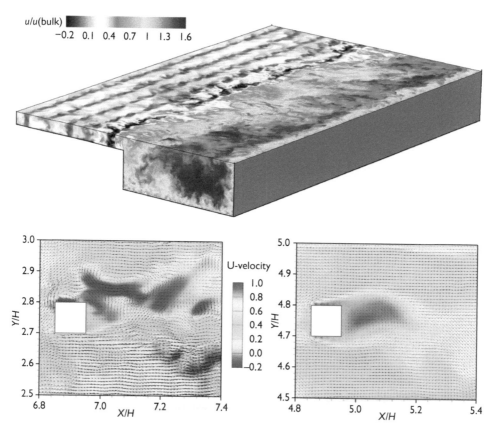

Figure 9.32 Flow in an asymmetric compound channel with a shallow floodplain roughened with ide-alized emergent vegetation (case 3): 3D view (top) and blow ups (bottom) of contours of instantaneous streamwise velocity. The lower left figure is a blow-up of the flow close to the interface, the lower right is a blow up of the flow further away from the interface. From Kara et al (2012b).

observed (bottom left part of Fig. 9.32), while further into the floodplain, i.e. away from the interface, vortices are shed regularly (bottom right part of Fig. 9.32).

Large-scale horizontal turbulence structures, vortex shedding and secondary flows lead to mass exchange across the main-channel-floodplain-interface. Figure 9.33 presents contours of normalized tracer concentration, which visualizes the afore-mentioned hor-izontal structures. At $x/H = 0$ the entire floodplain width is covered with the tracer; then the tracer is continuously mixed along the streamwise direction mainly by virtue of the turbulent structures. In the cases with floodplain vegetation, here only case 3 is shown, the structures are more consistent and are larger than in the smooth floodplain case. As a result, tracer is more efficiently mixed along the channel. This is better quantified in a time-averaged view of the concentration field as presented in Figure 9.34. In addi-tion to the horizontal interface vortices, the von Karman-type structures enhance the mixing of the tracer, and hence the mixing layer penetrates further onto the floodplain.

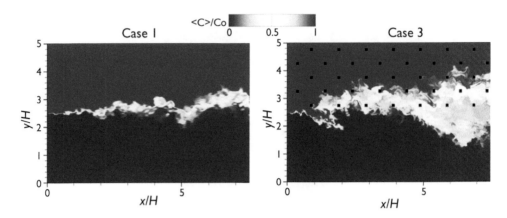

Figure 9.33 Flow in asymmetric compound channels: Distribution of instantaneous tracer concentration at the water surface for the smooth floodplain (Case 1) and the vegetated floodplain with $\phi = 0.04$ (Case 3). From Kara et al (2012b).

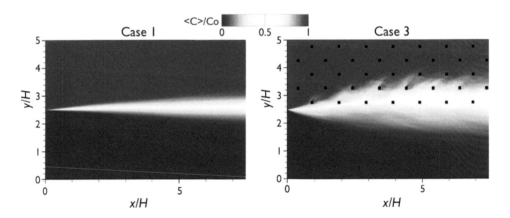

Figure 9.34 Flow in asymmetric compound channels: Distribution of time-averaged tracer concentration at the water surface for the smooth floodplain (Case 1) and the vegetation – roughened floodplain (Case 3). From Kara et al (2012b).

9.6 FLOW IN CURVED OPEN CHANNELS

The majority of rivers contain meandering regions in which channel curvature effects are important. The evolution in time of the river is controlled to a large extent by erosion at the bed and outer bank due to redistribution of the streamwise momentum caused by curvature-induced differences in the streamwise pressure gradients and secondary flow within and downstream of high-curvature reaches and by the secondary flow itself. Because of the redistribution of the streamwise velocity, in channels with bathymetries that are close to flat the core of high streamwise velocities moves first toward the inner bank and then toward the outer bank. In alluvial channels, the resulting bed shear stresses induce sediment erosion near the outer bank where a pool is created.

To first approximation, curvature effects and the strength of the secondary flow (e.g. as measured by the circulation of the main cell of cross-stream motion) vary monotonically with the ratio of the local curvature radius, R, to the channel width, B. Generally, the ratio R/B is used to classify bends as being of low ($R/B > 8$), medium ($8 > R/B > 3$) or high ($R/B < 3$) curvature. As R/B decreases, the degree of nonlinearity of the interactions between the secondary flow and the streamwise momentum increases. These interactions also increase the role of turbulence and modify sediment erosion and deposition patterns in loose-bed channels (Blanckaert and de Vriend, 2004).

LES with wall functions was used by van Balen et al. (2009) to study flow in a bend of medium curvature with a flat bed, by Stoesser et al. (2010b) to study flow in a S-shaped open channel with a flat bed (Re = 18,500) and by Moncho-Esteve et al. (2010) to predict flow in a compound meandering channel of medium curvature with flat bed and to understand the effects of the floodplain on the channel flow at flooding conditions. It was also shown earlier that RANS models with isotropic eddy-viscosity closures can predict fairly well the velocity redistribution and the main cell of cross-stream motion (Leschziner and Rodi, 1979, Zheng et al., 2008) but are less successful in predicting other details of the secondary flow and overestimate the friction losses (van Balen et al., 2010a). Well resolved LES of flow in a sharply curved 135° channel with deformed bathymetry ($R/B = 1.5$, Re = 60,000) were reported by Constantinescu et al. (2013a).

Detailed measurements of the mean flow in a 193° sharply curved channel ($R/B = 1.3$, Re = $UH/\nu \approx 68,000$, where H and U are the channel depth and bulk velocity in the upstream part of the inlet straight reach) with flat (case FB-LR) and deformed topography corresponding to equilibrium scour conditions (case DB-LR) were reported by Blanckaert (2010) and Zeng et al. (2008). These measurements constitute a unique set of data for understanding the flow physics in curved open channels and for testing the predictive capabilities of various turbulence modeling approaches. Both cases were simulated by van Balen et al. (2010a, 2010b) using LES with wall functions and the classical Smagorinsky model and by Constantinescu et al. (2011a, 2012a) using DES. Additionally, DES was used by Constantinescu et al. (2012a) to investigate Reynolds-number-induced scale effects in the same flat-bed bend, based on the comparison of simulations of case FB-LR and case FB-HR, for which Re = 10^6. To better bring out the predictive abilities of DES, results of RANS simulations with near-wall models are included for cases FB-LR and DB-LR (Zeng et al., 2008). The RANS calculations were performed with the Spalart-Allmaras (SA) and the k-ω SST models which gave very close results. This is why only one set is included.

In the SA-DES simulations discussed below, turbulent inflow conditions corresponding to fully-developed turbulent channel flow with resolved turbulent fluctuations were applied. A steady fully-developed precalculated RANS solution was used to specify the inflow conditions in the RANS simulations (Zeng et al., 2008). At the outflow, the convective boundary condition was used in DES. All the solid surfaces were treated as no-slip boundaries. In the RANS and DES of case DB-LR, the equivalent total bed roughness height estimated using the procedure described by Zeng et al. (2008) was 0.037 m. The implementation of rough-wall boundary conditions in the SA model is given in Spalart (2000). The free surface was treated as a rigid lid, which is justified as the channel Froude number is smaller than 0.4. The computational domain in the RANS and DES simulations (Fig. 9.35) was meshed using about 12 million cells for the LR cases and 18 million cells for the FB-HR case.

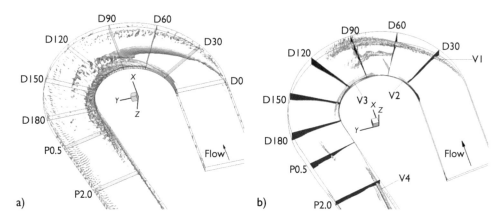

Figure 9.35 Flow in a curved open channel: Vortical structure of the mean flow (Q criterion) predicted by DES for: a) case FB-LR (from Constantinescu et al., 2012a); b) and DB-LR (from
Constantinescu et al., 2011a). V1 is the main cell of cross stream motion. Figure 9.35b is
reproduced by permission of the American Geophysical Union.

DES reveals that for both the flat and deformed-bed cases, besides the main cell of
cross-stream circulation (large vortex occupying most of the channel cross section in
section P0.5 in Figure 9.36 and the deeper part of the cross section in section D60 in
Figure 9.39), several Streamwise Oriented Vortical (SOV) cells form at the inner bank
within the bend (e.g., V2, V3 in Fig. 9.35b) and inside the downstream straight reach
(e.g., V4 in Fig. 9.35b). While in the flat-bed cases the axis of the main cell moves
from the inner bank toward the outer bank in the downstream part of the curved
reach (van Balen et al., 2009, 2010a), in the deformed-bed case the extent of the main
cell is generally limited to the deeper part of the section close to the outer bank (e.g.,
V1 in Fig. 9.35b). Because of the relatively high value of the Q isosurface needed to
visualize the SOV cells and the shear-layer regions in the high curvature bend, the
main cell V1 is not visible in Fig. 9.35a.

The region of high (positive) streamwise vorticity magnitude in sections D90 and
D180 (Fig. 9.36), which runs continuously from the outer wall to the inner wall close
to the channel bottom and then extends upwards parallel to the inner wall toward
the free surface, is induced by the main cell, V1. Compared to DES and experiment,
RANS strongly underestimates the thickness of this region and the vorticity levels
within it. The streamwise vorticity distributions predicted by DES in section D90 in
the flat bed cases (Fig. 9.36) show a patch of concentrated vorticity associated with
the SOV cell present close to the junction line between the inner left bank and the free
surface. This is consistent with the vorticity distributions obtained from experiment,
especially in section D90 where the measurement region extended close to the sidewalls. RANS could not capture this vortex.

A weak outer cell of cross-stream motion is present close to the outer bank in
the experiment (Fig. 9.36). DES of case FB-LR predicts a very weak outer cell whose
extent is limited to the upstream part of the bend. The DES of case FB-HR predicts the

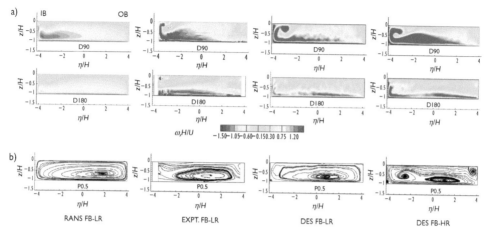

Figure 9.36 Flow in a curved open channel: a) Distribution of streamwise vorticity, $\omega_x H/U$, in sections D90 and D180; b) Secondary-flow streamlines in section P0.5 given by RANS, DES and experiment for case FB-LR (from Constantinescu et al., 2012a). Also shown are DES predictions for case FB-HR (courtesy of G. Constantinescu). IB and OB denote the inner bank and the outer bank, respectively. The transverse coordinate in each cross section is denoted η.

formation of an outer cell that extends not only within the curved reach but also over the upstream part of the straight outflow reach. This cell is visualized in Fig. 9.36 for section P0.5. Thus, an important Reynolds-number effect is the increase in coherence of the outer-bank cell. The non-dimensional circulation associated with the main cell increases by about 20–30% as a result of increasing the Reynolds number by about one order of magnitude. This scale effect can also be inferred from the comparison of the streamwise vorticity distributions for cases FB-LR and FB-HR in section D90 (Fig. 9.36) where the circulation of V1 reaches its peak.

The distributions of the bed-shear-stress magnitude nondimensionalized by its mean value in the inflow straight reach for each simulation, τ_0, are qualitatively similar for the two flat-bed cases (Fig. 9.37). However, τ/τ_0 is larger in case FB-HR within the two regions of large bed shear stress, one situated between sections D0 and D90 close to the inner bank and the other between sections P0.0 and P1.5 near the outer bank. The formation of these two regions is consistent with the switch of the core of high streamwise velocities from the inner bank in the upstream part of the bend to the outer bank around the bend exit in the flat-bed cases. The initial shift of the core of high streamwise velocity towards the inner bank around the entrance into the bend is due to the favourable streamwise pressure gradient near the inner bank and the adverse one near the outer bank in the upstream part of the bend. In the part approaching the exit and beyond, the pressure gradients are the other way around so that the velocity maximum shifts toward the outer bank, and this shift is also supported by the secondary motion (Leschziner and Rodi, 1979).

The curvature-induced interaction between the streamwise velocity and the cross-stream circulation in case DB-LR is complicated by the interaction with the pronounced riffle-pool bathymetry, which leaves a strong fingerprint on all characteristics

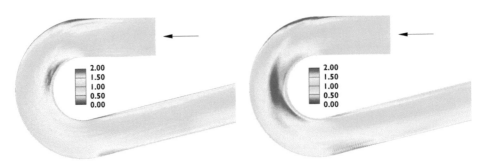

Figure 9.37 Flow in a curved open channel: Distribution of the nondimensional time-averaged bed shear stress magnitude, τ/τ_0, predicted by DES for case FB-LR (left) and case FB-HR (right) (from Constantinescu et al., 2012a).

of the flow field, including flow separation in horizontal planes and the associated formation of two recirculation zones over the shallowest regions at the inner bank (Fig. 9.38). The large vortical structure visible in Fig. 9.35b in the central part of the bend starting around section D60 is associated with the shear layer forming on the outside of the point bar and main recirculation region. The patch of high (positive) streamwise vorticity situated within the deeper (outer bank) side of sections D60 and D120 in Figure 9.39 is associated with the main cell, V1 (Fig. 9.35b). Judging from comparison with experiment, DES captures better the distribution of streamwise vorticity within the main cell compared to RANS in both sections.

A more accurate prediction of streamwise velocity and vorticity in the curved channel should result in more accurate predictions of the bed shear stress and the depth-averaged velocities that enter the sediment transport module in channel-morphodynamics calculations. While DES predicted the largest bed shear stresses to occur between sections D30 and D120 of the bend for case DB-LR (Fig. 9.37), RANS predicted the largest values to occur much farther downstream, between sections D70 and D180. The field study by Ferguson et al. (2003) of a channel with natural pool-riffle topography and strong inner-bank curvature ($R/B < 1.4$) found that the region of maximum boundary shear stress at the outer bank was situated mostly upstream of the bend apex, where a separated flow region similar to the one observed in case DB-LR was present, rather than downstream of it. This is consistent with the present DES results.

The flow through a natural-like meandering channel with a pool-riffle sequence (Fig. 9.40a) was investigated using LES and URANS by Kang and Sotiropoulos (2011a–b) for base and bankfull flow conditions corresponding to an experiment conducted at the St. Antony Falls Laboratory Outdoor Stream Facility. The wavelength of the meander was about 25 m and the channel width about 3.0 m. The mean channel depth varied between 0.1 m and 0.3 m for base and bankfull conditions. ADV velocity measurements were available for validation. The channel Reynolds numbers in the two test cases were about 2.4×10^5 and 10^5. The high-resolution scanned bathymetry that was used to generate the computational domain resolved all relevant bathymetry features down to 10–15 cm (Fig. 9.41b).

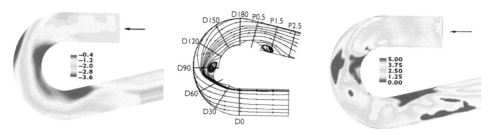

Figure 9.38 Flow in a curved open channel: Bathymetry (left), 2D streamline patterns at 0.5H below the free surface (middle) and nondimensional time-averaged bed shear stress, τ/τ_0 (right) predicted by DES in case DB-LR. The bed elevation (z/H) is measured with respect to the mean position of the free surface (z/H = 0) in the inlet section. From Constantinescu et al. (2011a); reproduced by permission of the American Geophysical Union.

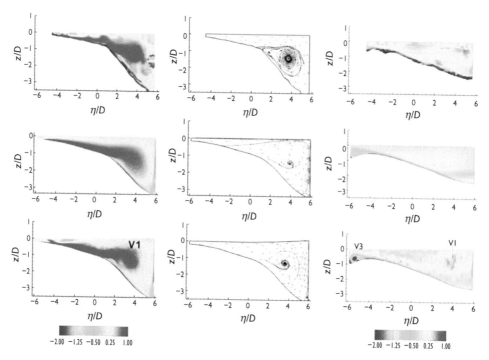

Figure 9.39 Flow in a curved open channel: Streamwise vorticity $\omega_x D/U$ (left) and secondary-flow streamline patterns (middle) in section D60 and streamwise vorticity in section D120 (right) obtained from experiment (top), RANS (middle) and DES (bottom) for case DB-LR. D is the flow depth in the incoming straight channel upstream of the region where erosion and deposition starts. From Constantinescu et al. (2011a), reproduced by permission of the American Geophysical Union.

The governing equations were discretized in curvilinear coordinates using a fractional-step algorithm. An immersed boundary method was used to describe the exact geometry of the channel. The computational domain in the bankfull case contained a total of 49 million cells of which 15 million were located in regions containing

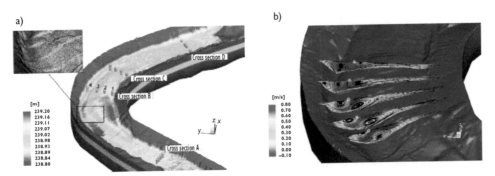

Figure 9.40 Flow in a meandering channel: a) Bathymetry of the meandering channel; b) Contours of mean streamwise velocity and secondary-flow streamline patterns at several cross sections near the bend apex. Results are shown for bankfull conditions. The contour levels in frame a) denote bed elevations. The flow direction is from bottom to top. From Kang and Sotiropoulos (2011), this material is reproduced with permission of John Wiley & Sons, Inc.

fluid. The LES simulations were performed using the dynamic Smagorinsky model and the simplified version of the wall model proposed by Wang and Moin (2002) in which the eddy viscosity is obtained using a mixing-length model with wall damping. Unsteady RANS simulations were performed using the SST model. The free surface was treated as a sloping rigid lid which is an acceptable approximation given the low value of the mean Froude number (Fr < 0.4). A precursor LES of flow in a periodic channel was used to generate the inflow conditions. The channel bed was treated as a smooth surface (i.e., the resolved small-scale bathymetry features are not roughness).

LES of the bankfull case (Kang and Sotiropoulos, 2011) revealed a very complex flow pattern with multiple SOV cells forming near the outer bank (Fig. 9.40b). The flow pattern was complicated by the rapid variation of the bathymetry close to the bank line and by the large-scale bathymetry features which strongly affected the near-bed flow. For example, some of the boulders present within the riffle region in Fig. 9.41b induced large-scale vortex shedding. Similar to the 193° bend case with equilibrium bathymetry (Fig. 9.38), the flow separated in horizontal planes close to the free surface over the point bar (Fig. 9.41a). Strong shear layers formed between the core of high streamwise velocity fluid and the slower moving fluid close to the banks. Energetic shear layers were also induced by the large-scale bank protrusions and bathymetry variations close to the bank lines. URANS simulations of the base flow case conducted using the SST model successfully reproduced most of the important features of the mean velocity and turbulent kinetic energy fields within the channel (Kang et al., 2011). Overall, LES showed a better level of agreement with experiment. In part, this was due to the known difficulty of RANS to resolve the details of the flow near the outer bank where turbulence-anisotropy effects strongly affect the pattern of secondary flow (Fig. 9.40b).

Results obtained for the test cases discussed here demonstrate that geometrically complex river-flow situations as they occur in nature can be calculated realistically by LES and DES. Besides a complex bathymetry, alluvial meandering channels can contain river training and other types of hydraulic structures. One of the most common measures to protect against severe erosion at the outer bank of curved alluvial channel

a)   b)

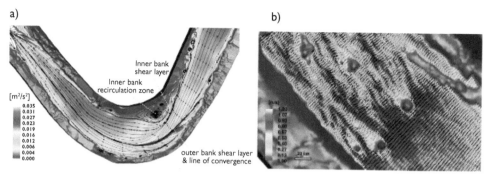

Figure 9.41 Flow in a meandering channel: a) Turbulent kinetic energy contours and streamline patterns at the free surface. The flow direction if from right to left; b) detail view of the instantaneous 2-D velocity vectors and velocity magnitude contours at a near-bed plane of the second riffle region showing the effect of the large-scale bed protrusions on the flow. Results are shown for bankfull conditions. From Kang and Sotiropoulos (2011), this material is reproduced with permission of John Wiley & Sons, Inc.

reaches is to install groynes. Recently, Kashyap et al. (2010) used LES without wall functions to predict flow in a high-curvature bend ($R/B \cong 1.5$, Re = 60,000) with a series of submerged groynes and deformed bed corresponding to high-flow conditions.

9.7 SHALLOW MERGING FLOWS

This section focuses on LES and DES of flow and transport processes in the Mixing Interface (MI) region downstream of two merging streams. Results are presented for two configurations. In the first one, the channel geometry is idealized and the incoming streams with different velocity are parallel. Downstream of the splitter plate separating these streams, a shallow Mixing Layer (ML) develops. The first configuration is relevant for understanding flow and transport processes at a river confluence in which the two incoming streams are close to parallel. The second configuration considers a natural river confluence with an angle between the two incoming streams close to 60°. Two cases with a low and, respectively, a high velocity ratio between the two streams are considered.

9.7.1 Shallow mixing layer developing between two parallel streams

Kirkil and Constantinescu (2008, 2009a) used DES to investigate the dynamics of a shallow ML developing between two parallel streams in a long open channel with both a flat smooth bottom and a bottom containing an array of identical 2D dunes. A detaied description of the SA-DES model and the numerical method are given in Constantinescu et al. (2003) and Constantinescu and Squires (2004). A splitter wall separates two developed turbulent channel flows (Fig. 9.42). Due to the presence of the free surface and the bed, the vertical development of the large-scale turbulent

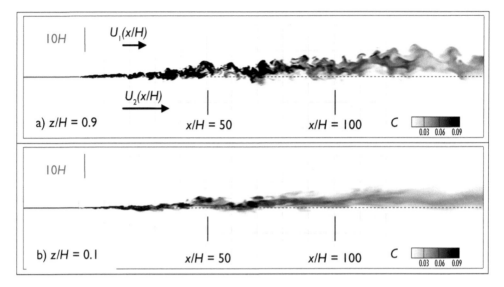

Figure 9.42 Shallow mixing layer between two parallel streams: Visualization of the structure of the shallow ML for the smooth-bed case using concentration contours. a) $z/D = 0.9$; b) $z/D = 0.1$. From Kirkil and Constantinescu, (2008).

structures in the shallow ML is constrained with respect to the widely studied case of a free ML. The growth of the large-scale quasi-2D horizontal ML eddies is driven by the transverse shear induced by the velocity difference between the two streams. The bed friction is the other main factor affecting the development of the ML. As discussed in Chu and Babarutsi (1988), a bed-friction number, S, can be introduced to characterize the stabilizing influence of the bottom friction on the ML development:

$$S = \frac{\bar{c}_f \delta}{2H} \frac{U}{\Delta U} \tag{9.1}$$

where H is the mean channel depth and $\bar{c}_f$ is the average bed friction coefficient of the two incoming streams. $\delta(x)$ is the ML width (definition given below), $U(x) = 0.5(U_1(x) + U_2(x))$ and $\Delta U(x) = (U_2(x) - U_1(x))$ are the mean velocity and the velocity difference across the ML. All these quantities are also a function of the distance z from the bed. The critical value of S is denoted S_c and corresponds to an equilibrium between stabilizing, and hence damping, effect of bed friction on the quasi 2D eddies that suppresses their growth and the destabilizing effect by the transverse shear on the same eddies. Experiments and stability analysis have shown that $0.06 < S_c < 0.12$ (Uijttewaal and Booij, 2000). Due to the velocity difference of the two incoming streams, the mean bed friction and thus the streamwise pressure gradients on the two sides of the ML are not equal. This induces at any given x location a transverse pressure gradient, which explains the shift of the ML centerline toward the low-speed side. As opposed to the case of a free ML, the streamwise growth of the ML width, $\delta(x)$, reduces with the distance from the splitter plate. Moreover, in the case of a channel of finite width, as a result of the shift of the centerline, $\Delta U(x)$ decreases in the streamwise direction. When $S = S_c$, the decrease in $U_2(x) - U_1(x)$ ceases and $\delta(x)$ becomes constant.

In the simulation discussed below, the Reynolds number $Re = U_0 H/\nu$, where U_0 (= 0.23 m/s) is the mean velocity of the two currents, was 15,400. These are the same conditions as in the smooth-bed experiments of Uijttewaal and Booij (2000) and van Prooijen and Uijttewaal (2002) for a channel depth $H = 0.067$ m. The width of the channel was $47H$ and its length downstream of the splitter plate was $157H$. The depth-averaged mean velocities of the incoming streams were $U_{10} = 0.61 U_0$ and $U_{20} = 1.39 U_0$, respectively. The mesh contained close to 10 million cells with 32 points in the vertical direction. The first grid point was situated at a distance of about one wall unit from the bottom wall and the sidewalls. The ML width was resolved using 20–60 grid points. The incoming flow contained realistic turbulence fluctuations obtained from two precursor LES of developed channel flow with bulk velocities U_{10} and U_{20}, respectively. Mass exchange processes were studied by considering the transport of a passive scalar introduced continuously at the end of the splitter plate over the whole depth of the channel (Fig. 9.42). The molecular and turbulent Schmidt numbers were assumed equal to unity. In the smooth-bed simulation the average value of the bed friction coefficient in the two incoming streams defined with the bulk velocity and averaged bed friction velocity was $\bar{c}_f = 0.5(\bar{c}_{f1} + \bar{c}_{f2}) = 0.0051$ where $\bar{c}_{fi} = 0.5(u_i^*/U_{i0})^2$ with $i = 1,2$. In the rough-bed simulation, an array of identical 2-D dunes was present at the bed. The wavelength of the dunes was $3.75H$ and their height was $0.25H$. The equivalent non-dimensional bed roughness height k_s^+ was close to 200 (fully-rough regime) and $\bar{c}_f \cong 0.015$. The following discussion focuses on the smooth-bed case.

The channel length was sufficiently large to allow the development of large-scale quasi-2D ML eddies whose diameter became much larger than H. Close to the free surface ($z/H = 1$), the average size of the largest eddies in the transverse direction was around $10H$–$15H$ in the downstream part of the channel ($x/H > 120$). The development of the ML was strongly non-uniform in the vertical direction. This can be inferred from comparison of the concentration fields at $z/H = 0.1$ and $z/H = 0.9$ in Figure 9.42. Figure 9.43 allows a more quantitative comparison of the ML width close to the bed ($z/H = 0.1$) and the free surface ($z/H = 0.9$). The ML width at a certain distance from the bed, z, is defined using the maximum slope thickness in the streamwise velocity, $\delta(x,z) = (U_1 - U_2)/(\partial u/\partial y)_{max}$. In the range $x/H = 75$–150, δ close to the free surface is 25–30% larger than in the near-bed region. The predicted δ-values at $z/H = 0.9$ are close to the measurements of van Prooijen and Uijttewaal (2002). The spanwise profiles of the non-dimensional mean streamwise velocity $(u - U_1)/(U_2 - U_1)$ at $z/H = 0.9$ in Figure 9.44 are close to self-similar and in good agreement with the ones measured by Uijttewaal and Booij (2000). The lateral coordinate is scaled with δ. The profiles are close to an error function.

Comparison of the instantaneous concentration fields at $z/H = 0.1$ and 0.9 (Fig. 9.42) shows that the coherence and size of the ML eddies decrease as the channel bottom is approached. There, a clear loss of coherence of the ML eddies is observed for $x/H > 100$. This phenomenon is similar to the case of open channel turbulence without mean shear where the flow structure close to the channel bottom is also dominated by small-scale bottom-generated turbulence (Fig. 9.3). The streamwise rate of growth of δ (Fig. 9.43) and the rate of growth of the ML centerline shift, s, (Fig. 9.45)

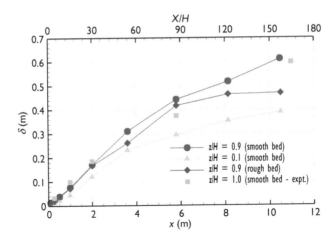

Figure 9.43 Shallow mixing layer between two parallel streams: Variation of ML width, δ, with stream-wise distance for cases with a smooth bed and a rough bed. Also shown is comparison with measurements of δ at the free surface (van Prooijen and Uijttewaal, 2002) for the case of a smooth bed. From Kirkil and Constantinescu, (2009a).

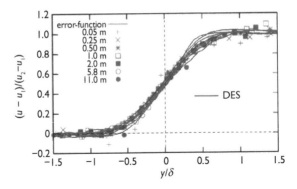

Figure 9.44 Shallow mixing layer between parallel streams: Lateral profiles of non-dimensional mean streamwise velocity at $z/D = 0.9$ for the smooth-bed case (from Kirkil and Constantinescu, 2008). The experimental data of Uijttewaal and Booij (2000) are shown using symbols.

decrease in this downstream region. The variations of $s(x)$ close to the free surface in the smooth-bed simulation and the corresponding experiment are similar. The increase in bed roughness can be seen to induce a much more rapid and larger shift of the ML centerline with respect to the smooth-bed case. Further, the ML width at a given streamwise location decreases with the increase in bed roughness (Fig. 9.43). Both these effects are due to increased differences in bottom friction between the two streams in the rough-bed case. As the bed roughness, and hence c_f increases, according to Equation (9.1) the bed friction number S is larger and reaches faster the critical value S_c so that the growth of δ reduces earlier.

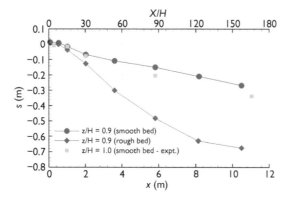

Figure 9.45 Shallow mixing layer between parallel streams: Variation of lateral shift of ML centerline, s, with streamwise distance at z/D = 0.9 (from Kirkil and Constantinescu, 2008). The experimental data of van Prooijen and Uijttewaal (2002) for the smooth-bed case are shown with square symbols.

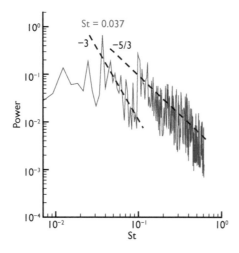

Figure 9.46 Shallow mixing layer between two parallel streams: Power spectrum of spanwise velocity fluctuations at a point close to the free surface (z/D = 0.9) located near the ML centerline and at 4.67 m from the splitter plate for the smooth bed case (from Kirkil and Constantinescu, 2009a).

For distances larger than $10H$ from the splitter plate, power spectra of the horizontal velocity components show a 2-range behaviour (as schematically sketched in Fig. 1.3) with a low-wave number subrange due to the large-scale quasi 2-D eddies and a higher-wave number range representing the small-scale 3-D turbulence. The latter is characterized by the usual $-5/3$ decay law. In the low wave-number subrange, where a -3 decay law prevails along with a transfer of energy from the smaller scales to the large 2-D eddies, the frequency peak of the spanwise velocity fluctuations measured by Uijttewaal and Booij (2000) at $x = 5.75$ m was 0.14 Hz, which corresponds to St $\cong$ 0.042. This value is in good agreement with the predicted frequency peak at

$x = 4.67$ m (St = 0.037, see Fig. 9.46). Additional analysis of the autocorrelation function of the spanwise velocity fluctuations close to the free surface showed that the average streamwise length of the quasi 2-D eddies is about 2.5 to 3.0 times larger than the local ML width (Kirkil and Constantinescu, 2009a). This is consistent with values inferred from experiment. It shows that the large-scale eddies are not circular, but rather elongated in the streamwise direction (Fig. 9.42).

9.7.2   River confluences

The flow and the turbulence structure at river confluences between non-parallel streams are characterized by the formation of a Mixing Interface (MI) and, in many cases, of Streamwise-Oriented Vortical (SOV) cells on the sides of the MI. The SOV cells form because of the transverse momentum of the two colliding streams. Depending on the angles between the two incoming streams and the downstream channel, and the velocity and momentum ratios between the incoming streams, the MI can be in the Kelvin-Helmholtz (KH) mode or in the wake mode, following the classification introduced by Constantinescu et al. (2011b). In the former case, the MI contains predominantly co-rotating large-scale (horizontal) quasi 2-D eddies whose growth is primarily driven by the KH instability and vortex pairing. The axes of these eddies are close to vertical. In the latter case, the MI is populated by quasi 2-D eddies with opposing senses of rotation that are shed from the junction corner region as a result of the interaction between the Separated Shear Layers (SSLs) on the two sides of the junction corner. Both types of MIs were observed in field investigations of the flow and turbulence structure at the confluence of the Kaskaskia River (KR) and Copper Slough (CS) stream in Illinois, USA conducted by Sukhodolov and Rhoads (2001) and Rhoads and Sukhodolov (2001). The cross sections of both tributaries upstream of the confluence are trapezoidal. The upstream channel for the Kaskaskia River is fairly well aligned with the downstream channel. The Copper Slough joins the Kaskaskia River at an angle of about 60°. The inner and the outer banks correspond to the east (E) and the west (W) banks, respectively.

The results discussed below are for two test cases of this river confluence investigated with DES by Constantinescu et al. (2011b, 2012b). These references provide a detailed discussion of the flow physics at river confluences with a large angle between the incoming streams. The SA-DES model used to perform the simulations is the same as the one used to perform the shallow ML simulations in section 9.7.1. The geometry of these test cases can be seen in Figures 9.47, 9.48 and 9.51b. The first test case (Case 1) was for a momentum ratio, Mr $\cong$ 1, and a Reynolds number based on mean values of velocity and flow depth in the downstream channel of Re = 166,000. The second (Case 2) was for Mr $\cong$ 5.5 and Re = 77,000. The mean flow depth in Case 2 was about two thirds of that in Case 1. For both test cases, every quantity was non-dimensionalized using the mean velocity (U) and the mean flow depth in the downstream channel, (H) in Case 1, i.e., $U = 0.45$ m/s and $H = 0.36$ m. The computational domain in the two test cases was meshed with close to 5 million cells. Based on information from the field studies, the channel bed and banks were treated as rough surfaces with a mean value of the bed roughness height of 0.01 m. The wall-normal grid spacing of the first row of cells off the bed and the banks was less than two wall units. Inflow conditions corresponding to fully-developed turbulent channel flow with resolved turbulent fluctuations were applied for the two incoming streams. The velocity fluctuations were obtained from LES conducted at a lower channel Reynolds number (Re $\cong$ 10,000), while the mean flow was obtained from RANS conducted with the flow conditions in the two incoming streams observed in the field experiments. A convective boundary condition was used at the outflow of the domain. The free surface was modeled as a shear-free rigid lid.

Though the horizontal eddies convected inside the MI were found to be quasi 2-D in both cases (energy spectra for the horizontal velocity components contained a subrange with a decay exponent of -3), the position of the MI (Fig. 9.47) and the mechanisms responsible for the formation and the dynamics of the quasi 2-D eddies were significantly different. In Case 1, the MI was situated within the central part of the downstream channel. In Case 2, the much larger value of the momentum of the CS stream and the presence of a very shallow region close to the east bank between sections A and C caused the MI to move toward the west bank. The position of the MI in both cases was in very good agreement with the one inferred from field data.

In Case 1 (Fig. 9.47a), the wake mode is dominant. This is why the growth in size of the quasi 2-D ML eddies was not accompanied by a noticeable increase in their circulation. Thus, their capacity to entrain sediment from the bed at large distances from the upstream junction corner is limited. In Case 2 (Fig. 9.47b), the KH mode is dominant. The increase in the mean size of the co-rotating quasi 2-D eddies as a result of vortex pairing events (e.g., these eddies attained a diameter of 7–8H by section C) was accompanied by a significant increase of their circulation. This explains why the capacity of the MI eddies to entrain sediment was significantly larger in Case 2 compared to Case 1.

Despite the fact that the position of the MI with respect to the two banks changed considerably, a system of strongly coherent SOV cells developed on the two sides (denoted E on CS side and W on KR side) of the MI in both cases (Fig. 9.48). The cores of the SOV cells are close to circular (Fig. 9.51b). In Case 1 (Mr $\cong$ 1), the circulations of the SOV cells on the two sides of the MI are within 50% of each other downstream

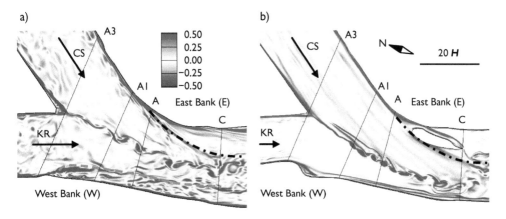

Figure 9.47 Flow at a river confluence: Distribution of instantaneous vertical vorticity, $\omega_z H/U$, in a horizontal surface 0.1H below the free surface. a) Case 1; b) Case 2. The black dash-dot line visualizes the shear layer forming at the east bank due to the strong curvature of the inner bank. From Constantinescu et al. (2012b); reproduced by permission of the American Geophysical Union.

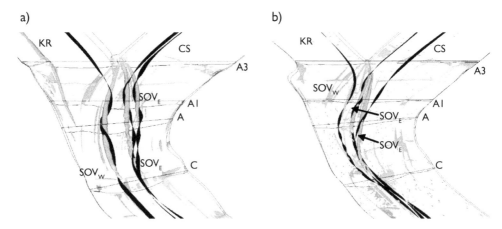

Figure 9.48 Flow at a river confluence: Visualization of the main vortical structures in the mean flow using a Q iso-surface. a) Case 1; b) Case 2. The 3-D ribbons visualize the helical motion of particles inside the SOV cells. From Constantinescu et al. (2012b), reproduced by permission of the American Geophysical Union.

of section A1. In Case 2 (Mr ≅ 5.5), the much larger momentum of the CS stream induces SOV cells of larger coherence and circulation on the CS side of the MI. The SOV cells on the KR side are quite weak in Case 2, even in the upstream region of the MI where the KR stream approaches the MI at a high angle. The ratio between the circulations of the SOV cells on the two sides of the MI is close to 3.5 between sections A3 and A. The study of Constantinescu et al. (2012b) suggests that strongly coherent SOV cells are a common characteristic of flow in confluences with concordant beds provided that the angle between one or both of the incoming tributaries and the downstream channel is large, which is generally the case for natural confluences.

DES also showed the critical role played by the SOV cells in the redistribution of the streamwise momentum on the two sides of the MI. For example, in Case 1 the distribution of the streamwise velocity field predicted by DES contains two distinct regions of large streamwise velocity u_s within the upper part of section A (Fig. 9.49b). The high-velocity region ($u_s > 0.5$ m/s) on the west side of section A is larger than on the east side, in good agreement with field measurements (Fig. 9.49a). The circulation of the SOV cell on the east side is larger than that on the west side, causing the core of high u_s values to be displaced toward the east bank in section A. By contrast, RANS does not predict two regions of relatively high u_s values close to the free surface (Fig. 9.49c). The main reason is that RANS severely underpredicts the coherence and circulation of the SOV cells (Constantinescu et al., 2011b).

In both cases, the core of the SOV cell on the side of the incoming stream making a large angle with the downstream channel was observed to undergo large-scale bimodal oscillations toward (interface mode, IM) and away (bank mode, BM) from the MI (e.g., see Fig. 9.50 for case 1 and detailed discussion in Constantinescu et al., 2012b). The presence of these oscillations explains why transverse velocity fluctuations in the field experiment with Mr ≅ 1 were comparable to or even larger than streamwise velocity fluctuations in the central part of the cross-sections downstream of

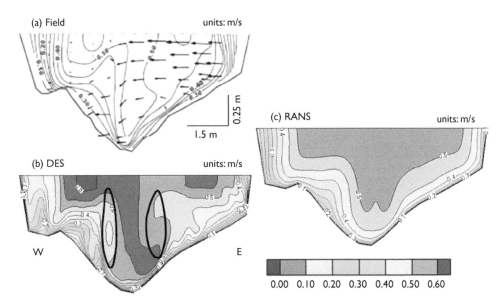

Figure 9.49 Flow at a river confluence: Distribution of the mean streamwise velocity, u_s (m/s), in section A for Case 1. a) field experiment; b) DES; c) RANS. The scale is distorted in the vertical direction (aspect ratio is 1:0.208). The two thick solid lines visualize the position of the main SOV cells in the section. From Constantinescu et al. (2011b); reproduced by permission of the American Geophysical Union.

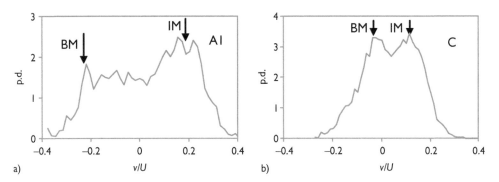

Figure 9.50 Flow at a river confluence: Probability-density functions of the vertical component of the instantaneous vertical velocity, v/U, for Case 1 at a point situated close to the axis of SOV_E in the mean flow. a) section A1; b) section C. IM and BM denote the interface mode and the bank mode, respectively. From Constantinescu et al. (2012b), reproduced by permission of the American Geophysical Union.

the junction corner. The bimodal oscillations were found to generate a strong local amplification of the turbulent kinetic energy. In both cases, the largest bed friction velocities were recorded beneath the region where the core of the primary SOV cell was subject to strong bimodal oscillations (e.g., see Fig. 9.51a for Case 1). In Case 2,

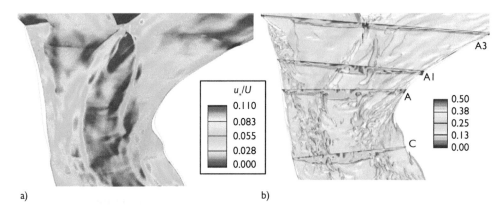

a)   b)

Figure 9.51 Flow at a river confluence: a) Distribution of the magnitude of the instantaneous bed
friction velocity, u_*/U, Case 1. b) vortical structure of the flow at the same time instant, as
visualized by the $\hat{Q}$ criterion and by vorticity magnitude, $\omega H/U$, at sections A3, A1, A and C.
Figure 9.51a, courtesy of G. Constantinescu. Figure 9.51b is from Constantinescu et al.
(2011b), reproduced by permission of the American Geophysical Union.

the bimodal oscillations were weaker than those observed in Case 1 and limited to
the region where the central part of the incoming, higher-momentum stream collided
with the MI.

DES also allowed to explain the main features of the distributions of the bed
friction velocity, u_*/U, on the basis of the large-scale structures present in the mean
and instantaneous flow fields. For example, for Case 1 the largest mean values of
u_*/U were observed beneath the primary SOV cells on the two sides of the MI. At
any given time instant, these SOV cells were characterized by a large non-uniform
variation of their coherence in the streamwise direction (e.g., see Fig. 9.51b for
Case 1). In the instantaneous flow fields, the core of the SOV cell on the east side
loses most of its coherence by section C. Downstream of section A, the predominantly streamwise-oriented eddies detaching from the downstream part of this
SOV cell are strongly stretched by the surrounding turbulence, thereby increasing
their three-dimensionality and creating conditions for their breakup into smaller
3-D eddies.

The distribution of u_*/U in the instantaneous flow fields (Fig. 9.51a) is mainly
affected by the presence of large-scale energetic eddies close to the bed. The cores of
the SOV cells are twisting in time around axes whose distances from the bed vary
significantly over relatively small streamwise distances (Fig. 9.51b). Thus, even where
the coherence of a SOV cell is high, the zone of amplification of u_*/U beneath the
SOV cell will be a strong function of the local distance between the axis of the SOV
cell and the stream bed. A main finding of the study of Constantinescu et al. (2012b)
is that for confluences with a large angle between the two incoming streams, the
capacity of the most coherent SOV cells to entrain sediment is larger than or at least
comparable to that of the eddies convected inside the MI, even for cases in which the
KH mode is very strong.

9.8   FLOW PAST IN-STREAM HYDRAULIC STRUCTURES

This chapter describes LES and hybrid RANS-LES calculations of flow around some common types of in-stream hydraulic structures (bridge piers and abutments, spur dikes and groynes). Besides investigating the mean flow and turbulence structure and the role played by the large-scale coherent structures, some of these studies also considered mass transport. Most of these studies focused on conditions corresponding to the start of the erosion and deposition process, when the hydraulic structure is placed in an open channel with a flat bed. However, some of the calculations were conducted with a deformed bed corresponding to equilibrium scour conditions, which allowed analyzing scour processes after a large scour hole has developed around the obstacle. Most of the simulations were conducted in straight channels with vertical non-erodible sidewalls. The free surface was simulated as a stress-free rigid lid.

9.8.1   Flow past bridge piers

Though here the focus will be on discussing the erosion potential of the main types of large-scale coherent structures (e.g., horseshoe vortices, vortex tubes inside the separated shear layers, shed vortices in the wake), several other aspects of flow past surface-mounted cylinders are important for hydraulic engineering. For example, the structure and spatial extent of the wake, the nature of the wake vortex shedding, the interaction between the horseshoe vortices and the separated shear layers (e.g., see Kirkil and Constantinescu, 2012) and the unsteady forces on the cylinders are of great importance in applications where the cylinders correspond to plant stems in a vegetated channel or to large-scale roughness elements in a channel.

Kirkil et al. (2006) used LES to study flow and scour mechanisms around a circular pier in a straight flat-bed open channel. The simulation used a non-dissipative LES solver (Mahesh et al., 2004) and a dynamic Smagorinsky model. The channel depth was $H = 1.12D$, where D is the diameter of the pier. The unstructured mesh used to simulate the flow at $Re_D = 16,000$ contained over 4 million cells. No wall functions were used. The inflow conditions containing resolved turbulent fluctuations were generated by a precursor LES of developed channel flow. Figure 9.52 visualizes the horseshoe vortex (HV) system in the mean flow. The HV region is characterized by a very large amplification of the turbulence in the region of the main horseshoe vortex (e.g., see distribution of pressure fluctuations $\overline{p'^2}$ in Fig. 9.53a. As shown first by Devenport and Simpson (1990), the reason for this large amplification is the fact that the core of the main horseshoe vortex, PV, is subject to low-frequency bimodal oscillations toward and away from the cylinder. The double-peaked velocity histogram in Fig. 9.53b confirms the bimodal nature of these oscillations. Figure 9.54a shows the HV system in the backflow mode when the core of the PV is larger and its shape is elliptical, while when the HV system is in the zero-flow mode (Fig. 9.54b) the core of PV is smaller and fairly circular. Kirkil (2007) has shown that unsteady RANS cannot reproduce these bimodal oscillations. This underpins the superiority of LES in capturing the unsteady dynamics of the HV system observed in experiments. The large-scale sweeping motions of the PV toward and away from the base of the pier enhance considerably its capacity to entrain sediment. Consistent with experimental

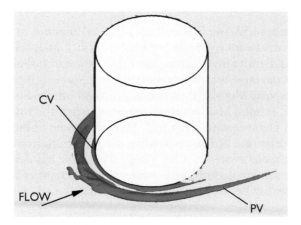

Figure 9.52 Flow past a circular bridge pier on a flat bed: Visualization of the mean flow structure of the HV system at the base of a circular pier (LES, $Re_D = 16,000$) using a Q-isosurface (from Kirkil et al., 2006).

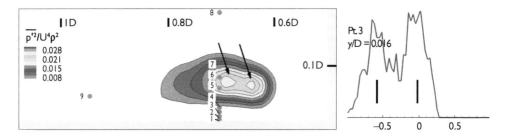

Figure 9.53 Flow past a circular bridge pier on a flat bed: Turbulence structure inside the horseshoe vortex system (LES, $Re_D = 16,000$). a) pressure fluctuations, $\overline{p'^2}/p^2U^4$, in the symmetry plane; b) velocity histogram for point 3 situated inside the region where the main horseshoe vortex is subject to bimodal oscillations (from Kirkil et al., 2006).

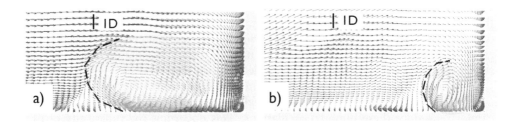

Figure 9.54 Flow past a circular bridge pier on a flat bed: Instantaneous velocity vectors in the symmetry plane revealing the structure of the flow inside the horseshoe vortex system (LES, $Re_D = 16,000$): a) backflow mode; b) zero-flow mode (from Kirkil, 2007).

observations showing that scour is initiated in the regions of strong flow acceleration around piers, the largest values of instantaneous u_* (Fig. 9.55) are observed close to the upstream face of the pier for polar angles between 30 and 80 degrees. Figure 9.55 also shows that the vortex tubes advected within the Separated Shear Layers (SSLs) and the wake roller vortices have a high capacity to entrain sediment.

While the horseshoe vortices play a dominant role in the growth of the scour hole in front of the pier, the formation and advection of large-scale shed (roller) vortices in the wake plays an important role for scouring behind the pier and in the transport of sediment away from the scoured region. In most cases, the vortex shedding behind the pier resembles the one observed for infinitely long cylinders which is dominated by the alternate shedding and advection of vortices as a result of the interaction between the SSLs on the two sides of the cylinder. The von Karman vortex street behind a circular pier ($Re_D = 9,030$) placed in a relatively shallow channel ($D = 1.66H$) predicted by a 3-D LES (Hinterberger et al., 2007) is visualized in Fig. 9.56b using a passive tracer injected near the stagnation line at the cylinder front, for which an advection-diffusion equation was solved. Comparison with experiment in Fig. 9.56a shows that the undulatory shape of the wake as well as the transport and diffusion of the roller vortices are very well captured by LES (see also animation at http://goo.gl/YXWKm). The dominant shedding frequency and the distributions of the mean streamwise mean velocity and velocity r.m.s. fluctuations in the wake were also found to be in good agreement with experiment. The simulation used a constant-coefficient Smagorinsky model and was conducted on a mesh containing 1.3 million cells.

Next, the effects of the shape of the pier are discussed by comparing results of DES simulations conducted by Kirkil and Constantinescu (2009b) at $Re_D = 2.4 \times 10^5$ for a circular pier and a high-aspect-ratio rectangular plate. The projected width of the two obstacles was the same ($W = D$). The mesh in the SA-DES contained around 8 million cells. The mean streamwise inflow velocity profile was obtained from a RANS simulation of fully developed channel flow and the fluctuations from an LES of such flow at a lower Reynolds number ($Re_D = 16.000$). The plate case is representative

Figure 9.55 Flow past a circular bridge pier on a flat bed: Distributions of bed friction velocity, u_*/U, around a circular pier (LES, $Re_D = 16,000$) at a time instant when the HV system is in the backflow mode (left) and in the zero flow mode (right). The values below the threshold value for sediment entrainment ($d_{50} = 0.68$ mm) were blanked out. The dashed circle and the arrow point toward the region of amplification of u_* induced by the formation of a wake roller (from Kirkil, 2007).

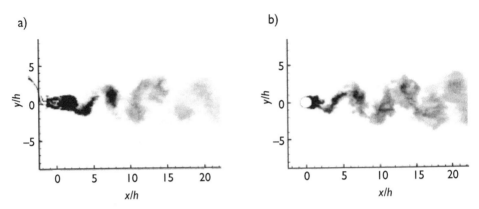

Figure 9.56 Flow past a circular bridge pier on a flat bed: Visualization of the wake past a circular pier with a flat bed (Re = 9,030) using a passive tracer injected left and right of the stagnation line at the front of the pier. a) experiment; b) LES. From Hinterberger et al. (2007), reprinted with permission from ASCE.

of piers with sharp edges (e.g., of rectangular shape) where the position of the boundary layer separation is fixed at all flow depths.

The structure of the HV system and the strength of the bimodal oscillations depend mainly on the bluntness of the frontal face of the in-stream obstacle and hence on its shape and orientation (Chang et al., 2011). DES investigations of the structure of the flow in the HV region and sediment entrainment mechanisms at high Reynolds numbers for different cylinder geometries (circular, wing-shaped piers and rectangular) were also reported by Escauriaza and Sotiropoulos (2011a), Paik et al. (2007) and Chang et al., (2011). The simulations conducted at $Re_D = 2.4 \times 10^5$ have shown that the circulation of the main horseshoe vortex and the level of amplification of the turbulence inside the HV region were significantly (up to 5–7 times) higher for the plate compared to the circular pier.

Similar to the LES of Hinterberger et al. (2007), the von Karman vortex street is also present in the DES conducted at a much larger Reynolds number (e.g., see Fig. 9.57 for the plate case). The flow shallowness and shape of the flow obstruction can affect the dominant frequency of the shed vortices, f, compared to the case of an infinitely long circular cylinder. In the latter case, the Strouhal number $St = fD/U$ is about 0.2 for Reynolds numbers below the drag crisis. The values inferred from the DES of the circular pier and plate were $St = 0.27$ and $St = 0.18$, respectively. Both values are in excellent agreement with the ones inferred from experiment (Kirkil and Constantinescu, 2009b).

More important for the wake scour processes is the fact that the coherence of the roller vortices was much larger in the case of the sharp-edged obstacle and remained large even in the near-bed region. The width of the wake was also slightly larger compared to the case of a circular pier. The important role of the wake vortices in entraining sediment behind piers with sharp edges is illustrated in Figure 9.58. The instantaneous values of u_* beneath the wake vortices are comparable to those observed in the region of high flow acceleration at the sides of the plate. The vortices can induce large values of u_* at large distances from the plate, which is not the case for circular

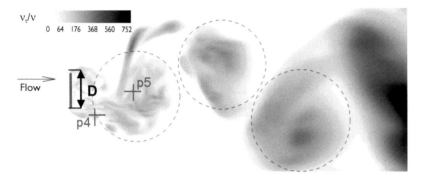

v_t/v
0 64 176 368 560 752

Flow

D

p5

p4

Figure 9.57 Flow past a surface-mounted rectangular plate on a flat bed: Visualization of the structure of the wake using eddy viscosity contours (DES, $Re_D = 2.4 \times 10^5$). The width of the obstacle was D, the same as the one in the circular cylinder case (from Kirkil, 2007).

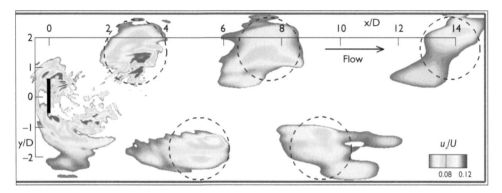

Figure 9.58 Flow past a surface-mounted rectangular plate on a flat bed: Distribution of instantaneous bed-friction velocity (DES, $Re_D = 2.4 \times 10^5$). The regions where $u_* < u_{*c0}$ ($d_{50} = 1.05$ mm) were blanked out. From Kirkil and Constantinescu (2009b), reproduced by permission of the American Geophysical Union.

piers where the roller vortices induced high values of u_* only in their formation region immediately behind the pier (e.g., see Fig. 9.55). This explains why in experiments the growth of the scour hole during the initial stages of the scour process is significantly faster for rectangular piers compared to circular piers having the same width.

Details of the mechanisms through which sediment particles are entrained from the bed and then transported past the pier can be obtained by coupling the Eulerian simulation of the flow fields with a Lagrangian particle model which uses discrete point particles. This approach was adopted by Escauriaza and Sotiropoulos (2011b) who simulated the flow and particle transport around a circular pier on a fixed flat bed at $Re_D = 39,000$. The set up corresponded to the experiments conducted by Dargahi (1989). The DES used a steady inflow profile obtained from a precursor RANS simulation. The mesh contained about 3 million cells. The number of point particles used in the simulations was 10^5, which is sufficient for simulating sediment transport under clear-water scour conditions. The particles were initially situated upstream of the cylinder. The results showed that their ejection from the bottom boundary layer, movement and trajectories were controlled to a large extent by the unsteady dynamics of the horseshoe vortices (Fig. 9.59). Two main modes of transport were identified: saltation and sliding along the bed. Simulation results also showed that a substantial number of particles were entrained along the legs of the horseshoe vortices.

It is only after the scour hole starts developing that the largest values of u_* will be consistently observed beneath the horseshoe vortices. At this point the HV system becomes the main mechanism responsible for the growth of the scour hole in front of the pier, where the most severe scour is observed. This is why it is important to understand the scour processes around piers which have already a large scour hole. Kirkil et al. (2008) used LES to study the flow structure around a circular pier with deformed bathymetry corresponding to equilibrium scour (Fig. 9.60a). The numerical method, mesh and boundary conditions were identical to those used by Kirkil et al., (2006). The bathymetry was obtained from an experiment conducted with a loose bed for clearwater scour conditions at $Re_D = 16,000$. The complex structure of the HV system in the mean flow is visualized in Figure 9.60b. Besides the primary horseshoe vortex HV1, the HV system contains a secondary co-rotating vortex, a small junction vortex, JV, and a bottom-attached horseshoe vortex, BAV, forming due to the interaction of HV1 with the deformed bed. Analysis of the instantaneous flow fields showed that, similar to the case of a flat bed, HV1 is subject to bimodal oscillations that result in a large amplification of the turbulence within the upstream part of the scour hole. Figure 9.61 visualizes the flow structure in front of the pier at two time instances when HV1 is in the backflow mode (Fig. 9.61a) and in the zero-flow mode (Fig. 9.61b). Similar to the flat bed case (Fig. 9.54), the injection of a patch of high-momentum fluid into the downflow results in the formation of a strong near-wall jet

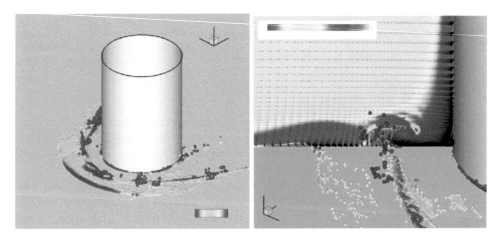

Figure 9.59 Flow past a circular bridge pier on a flat bed: Instantaneous particle transport with 10^5 particles ($d_{50} = 0.36$ mm) initially located in front of the pier ($Re_D = 39,000$). 3D view (left) and zoom around the main horseshoe vortex showing entrainment of particles into its core (right). From Escauriaza and Sotiropoulos, (2011b), reproduced with permission from Cambridge University Press.

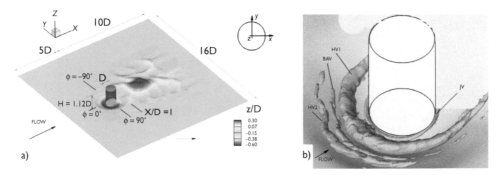

Figure 9.60 Flow past a circular bridge pier with deformed equilibrium bathymetry: a) computational domain and bathymetry; z/D = 0 corresponds to the bed level in the regions with no scour; b) visualization of the mean-flow horseshoe vortices inside the scour hole (LES, Re$_D$ = 16,000) using a Q isosurface. From Kirkil et al. (2008), reproduced with permission from ASCE.

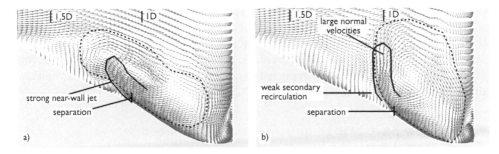

Figure 9.61 Flow past a circular bridge pier with deformed equilibrium bathymetry: Visualization of the structure of the flow inside the HV system in the symmetry plane (LES, Re$_D$ = 16,000) when the HV system is in: a) the backflow mode; b) the zero-flow mode. From Kirkil et al. (2008), reproduced with permission from ASCE.

parallel to the deformed bed that pushes HV1 away from the pier and triggers transition to the backflow mode. The bimodal oscillations are also the reason why the largest mean values of u_* are observed beneath HV1 (Fig. 9.62a).

As the bathymetry approaches equilibrium, the slope of the scour hole beneath the main horseshoe vortices increases. Gravitational bed-slope effects act toward reducing the capacity of the horseshoe vortices to entrain sediment, as the direction of their rotation is such that they try to move sediment particles against the local bed slope. In order to determine whether the predicted distribution of u_* is consistent with the equilibrium conditions present in the experiment, one has to check that $u_* < u_{*c}$. Here u_{*c} is the local value of the critical bed friction velocity corrected for gravitational bed slope effects as determined from the formula of Brooks and Shukry (1963). This value is different from the critical entrainment value $u_{*c0} = 0.106U$, which one would obtain from the Shields diagram for the case of sediment with $d_{50} = 0.68$ mm assuming a flat bed. Figure 9.62a shows that $u_*/u_{*c0} > 1$ over about 40% of the scoured region in front of the pier so that scour would continue if the bed were flat. However, the

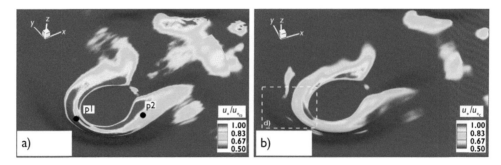

Figure 9.62 Flow past a circular bridge pier with deformed equilibrium bathymetry: Time-averaged bed-friction velocity contours (LES, Re_D = 16,000). a) u_*/u_{*c0}; b) u_*/u_{*c}. Red regions in frames a and b correspond to locations where u_*/u_{*c0} > 1 and u_*/u_{*c} > 1, respectively. From Kirkil et al. (2008), reproduced with permission from ASCE.

distribution of u_*/u_{*c} plotted in Figure 9.62b shows that u_*/u_{*c} < 1 over more than 97% of the scoured surface, so that indeed further scour does not occur and equilibrium prevails. Similar investigations of scour processes at much higher Reynolds numbers ($Re_D = 2.4 \times 10^5$) around circular piers and rectangular plates with flat and equilibrium deformed bed were reported by Kirkil et al. (2009), Kirkil and Constantinescu (2010) and Chang et al. (2011).

9.8.2 Flow past bridge abutments and isolated spur dikes

In this section results are shown from the LES and DES of Koken and Constantinescu (2008a, 2009a) of the flow around a vertical-wall abutment/spur dike placed in an open channel with a flat bed. The channel Reynolds numbers were 18,000 for LES (case LR) and 500,000 (case HR) for DES. The LES mesh contained 4 million cells while the DES mesh contained 7.5 million cells. In both simulations, the grid was fine enough to avoid the use of wall functions. The dynamic Smagorinsky SGS model was employed in LES while the SA model was used as the base model in DES. The inflow conditions were generated by a precursor LES of developed channel flow. The following discussion focuses on case HR which is more relevant for practical applications in hydraulics. Results for case LR are discussed in the context of Reynolds-number-induced scale effects.

In both cases, the main vortical systems in the flow are: 1) the horseshoe vortex (HV) system near the upstream base of the abutment; 2) the recirculating flow region upstream of the abutment; 3) the separated shear layer (SSL) originating at the tip of the abutment (see Figs. 9.65 and 9.66); 4) the recirculation region behind the abutment (Fig. 9.63); and 5) a streamwise-oriented vortex close to the sidewall (vortex VA in Fig. 9.63). As shown in Figure 9.63, the horseshoe vortices forming the HV system follow the junction line between the bed and the upstream face of the abutment. Past the abutment, the main horseshoe vortex, HV1, and the secondary horseshoe vortices are stretched and bent in the direction of the incoming flow. The role of the main corner vortex, CV1, is to convect fluid and momentum from the upper levels of the upstream recirculation region into the core of HV1. The main scale effect is that HV1 merges with the main secondary horseshoe vortex HV2 in case HR. The two vortices remain separated and the circulation of HV2 is smaller in case LR.

The structure and dynamics of the HV system in the two simulations were found to be qualitatively similar. The coherence of HV1 in time was modulated by its interaction with secondary horseshoe vortices and the bottom boundary layer. HV1 undergoes bimodal aperiodic oscillations, similar to the ones observed in experimental investigations of other junction flows (e.g., see discussion in section 9.8.1). Figure 9.64 shows the instantaneous velocity vectors in a plane cutting through the tip of the abutment at two time instants when HV1 is in the zero-flow mode and the backflow mode, respectively, for case HR. The average amplitude of the oscillations was smaller in case HR (Koken and Constantinescu, 2009a). In both cases, the intensity of the bimodal

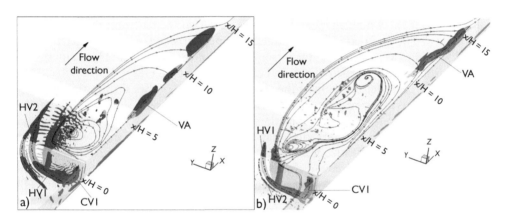

Figure 9.63 Flow past a bridge abutment: Vortical structure of the mean flow close to the abutment at the start of the scour process (flat bed). The horseshoe vortices HV1 and HV2 and the junction vortex VA are visualized using the Q criterion. Also shown are the 2-D streamline patterns at the free surface. a) case HR (from Koken and Constantinescu, 2009a); b) case LR. Figure 9.63a is reprinted with permission from Koken and Constantinescu (2009a). Copyright 2009, American Institute of Physics. Figure 9.63b is reproduced from Koken and Constantinescu (2008a) by permission of the American Geophysical Union.

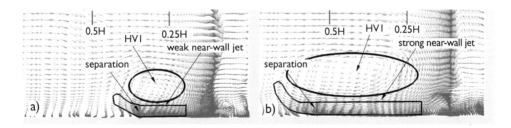

Figure 9.64 Flow past a bridge abutment: Instantaneous velocity vectors in a plane cutting through the tip of the abutment (case HR) corresponding to: a) the zero flow mode; b) the backflow mode of the HV system. Figure is reprinted with permission from Koken and Constantinescu (2009a). Copyright 2009, American Institute of Physics.

oscillations peaked at sections cutting through the tip of the abutment, where the circulation of HV1 and the overall amplification of the turbulence were the highest. The bimodal oscillations gradually disappeared in the leg of HV1. As the Reynolds number increased, HV1 became more stable and the size of its core decreased.

In the mean flow, the sizes of the main recirculation region downstream of the obstruction in the LR and HR simulations are very close (Fig. 9.63). The SSL reattaches at $x/H \cong 13$ (H is the channel depth). The maximum width of the recirculation region is 20% larger in the LR simulation. The flow inside the core of vortex VA forming along the sidewall moves in the upstream direction for $x/H < 13$, and in the opposite direction for $x/H > 13$. This feature of the flow inside vortex VA was confirmed by dye visualization experiments conducted by Koken and Constantinescu (2008a) for case LR. In the case of a loose bed, vortex VA plays an important role in the formation of the elongated deposition hill of sediment close to the sidewall (Koken and Constantinescu, 2008b), which is a general feature of a local scour bathymetry around a vertical wall abutment. In both cases, a wide range of energetic scales is observed over the whole channel depth inside the two recirculation regions forming upstream and downstream of the abutment and inside the SSL where energetic vortex-tube-like eddies are convected away from the tip of the abutment. This is illustrated by the instantaneous vorticity fields in Figure 9.65 for case HR.

The degree of deformation of the cores of the vortex tubes shed in the upstream part of the SSL is much larger in case HR (Fig. 9.66a) compared to case LR (Fig. 9.66b), in particular over the lower part of the channel ($z/H < 0.3$). Close to the bed, the vortex tubes reorient in a direction that makes a relatively low angle with the bed, thereby amplifying the horizontal vorticity in the near-bed region. By comparison, the cores of the vortex tubes remain close to vertical in case LR (Fig. 9.66b). Large-scale hairpin-like eddies are observed inside the downstream part of the SSL in case HR (Fig. 9.66a). The lower leg of some of these hairpin eddies is situated, at times, at a small distance from the bed and is able to strongly amplify the bed-friction velocity. Consequently, in case HR the instantaneous bed-friction velocity distribution displays a streaky structure over part of the SSL region (Fig. 9.67a), not observed in case LR (Fig. 9.67b). The amplification of u_* inside the streaks can be important, up to two times the mean value of u_* at the same location, as exemplified in Figure 9.68.

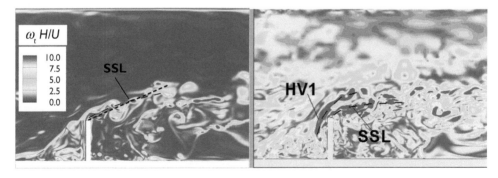

Figure 9.65 Flow past a bridge abutment: Distribution of the instantaneous vorticity magnitude for case HR. a) $z/H = 1$ (free surface); b) $z/H = 0.1$. Figure 9.63a is reprinted with permission from Koken and Constantinescu (2009a). Copyright 2009, American Institute of Physics.

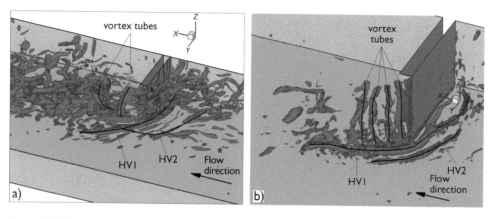

Figure 9.66 Flow past a bridge abutment: Visualization of vortical structure of the instantaneous flow using the Q criterion. a) case HR; b) case LR. Figure 9.66a is reprinted with permission from Koken and Constantinescu (2009a). Copyright 2009, American Institute of Physics. Frame 9.66b is reproduced from Koken and Constantinescu (2008a) by permission of the American Geophysical Union.

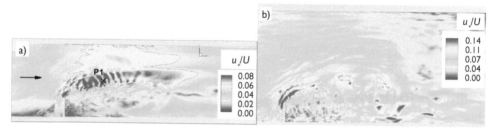

Figure 9.67 Flow past a bridge abutment: Distribution of the instantaneous bed-friction velocity. a) case HR; b) case LR. Figure is reprinted with permission from Koken and Constantinescu (2009a). Copyright 2009, American Institute of Physics.

The largest values of the time-averaged bed-friction velocity occur beneath the upstream part of the SSL and the main horseshoe vortex (Fig. 9.67). The large bed friction in the region around the tip of the abutment is mainly due to the acceleration of the flow as it passes around the abutment, rather than being induced by HV1. In case HR, a region of relatively high u_* is present between the outward side of the SSL and the sidewall opposite the abutment. The high u_* is caused by the increase in mean streamwise velocity resulting from the reduction of the effective flow area by the abutment (contraction scour effects). The ratio between the channel width and the abutment length was larger in case LR. This is the reason why in this case u_* remained relatively low outside the SSL.

Koken and Constantinescu (2008b, 2011a) discuss similar LES (Re = 18,000) and DES (Re = 240,000) investigations of flow past a vertical wall abutment/spur dike with equilibrium bathymetry obtained from experiment. Paik and Sotiropoulos (2005) used DES to simulate the flow past a rectangular abutment mounted on a non-erodible flat bed at Re = 420,000. Koken and Constantinescu (2009b, 2011b) discuss

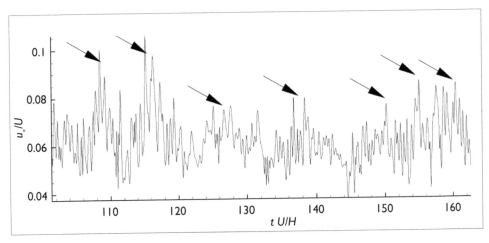

Figure 9.68 Flow past a bridge abutment: Time series of bed-friction velocity at point P1 (see Fig. 9.67a) for case HR. The arrows show the relative maxima that are induced by the passage of energetic SSL eddies (e.g., lower legs of hairpin-like vortices) at a small distance from the bed. Figure is reprinted with permission from Koken and Constantinescu (2009a). Copyright 2009, American Institute of Physics.

the changes in the flow structure and sediment entrainment mechanisms around abutments of more realistic geometry with sloped lateral walls with respect to the case of a vertical wall abutment. Their DES study considered both flat bed and equilibrium-bathymetry configurations. The sloped shape of the abutment deflects part of the incoming flow toward the upstream face of the abutment. This decreases the adverse pressure gradients and thus the coherence of the HV system vis à vis the case of a vertical-wall abutment.

9.8.3   Flow past groyne fields

This section discusses flow past groyne fields in a straight channel. Apart from their function as river control structures, the dead water zones within the embayments have a significant effect on the transport of pollutants in rivers (e.g., by increasing the longitudinal dispersion of pollutant clouds). The first case, referred to as Case 1, focuses on the flow and mass exchange processes around an embayment situated far from the start of the multiple-embayment groyne field region. The rank of the embayment denotes the embayment in the series, starting the count from the most upstream embayment. As the flow and mass exchange processes become independent of the rank of the embayment in the series, the flow over only one embayment can be calculated using periodic boundary conditions in the streamwise direction. Case 2 focuses on understanding the initial spatial development of the flow and mass exchange processes as a function of the rank of the embayment. The geometry of the embayments is different in the two test cases. The most important difference is that the channel depth was larger than the mean flow depth inside the embayments in Case 2. The channel depth was equal to the embayment depth in Case 1.

Case 1 was studied by Hinterberger et al. (2007) who used LES to simulate flow and tracer concentration in a groyne field that was investigated experimentally by Weitbrecht et al. (2008). The latter authors placed a series of 15 groynes in a flume and measured the surface flow by PIV and also the temporal development of passive scalar (colored dye) concentration in one downstream embayment initially seeded uniformly by the scalar. Because of periodicity of the flow, only the region corresponding to one embayment was computed in LES. The computation domain extended from the middle of one embayment to the middle of the next one. After the velocity field reached statistically steady state, the passive scalar concentration was set equal to 1 inside the embayment and equal to 0 in the main channel. The temporal development of the scalar field was calculated by solving the filtered advection-diffusion equation with a turbulent Schmidt number of 0.7.

The ratio of width W to length L of the embayment was $W/L = 0.4$, the ratio of W to water depth was $W/H = 10.8$ and the channel Reynolds number was Re = 7,340. The standard Smagorinsky SGS model was used. As the grid having 3 million cells was not fine enough for wall-resolving LES, wall functions were employed at all walls. Free slip conditions were used at the lateral boundary opposite the one containing the groynes. For $W/L = 0.4 < 0.5$, a two-gyre circulation pattern is expected (Uijttewaal et al., 2001). The main recirculation flow in the groyne field was well predicted, but the secondary recirculation in the lee of the upstream groyne was underpredicted compared with experiment, particularly near the surface.

The instantaneous flow field exhibiting vortices in the interfacial mixing layer region was simulated in good agreement with the experiment, as can be seen from comparison of the snapshots in Figures 9.69a (LES) and 9.69b (experiment). These vortices can also be seen clearly in Figure 9.69c which is a snapshot of contours of the

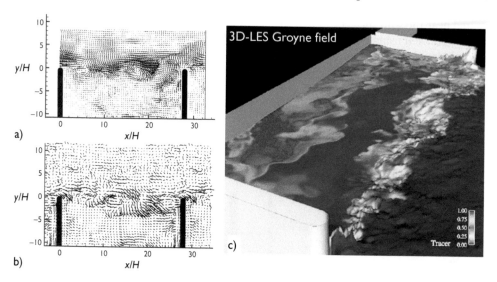

Figure 9.69 Flow and mass exchange between a channel and a groyne field: Visualization of the instantaneous flow, coherent structures and mass exchange at the free surface around an embayment for Case 1. a) instantaneous perturbation velocity vectors, PIV experiment; b) instantaneous perturbation velocity vectors, LES; c) instantaneous scalar concentration (courtesy of C. Hinterberger). Figures 9.69a–b are from Hinterberger et al. (2007), reprinted with permission from ASCE.

surface concentration and also of the surface elevation calculated from the pressure distribution at the frictionless rigid lid representing the free surface. The movie (http://goo.gl/nPbeA) from which this snapshot is taken illustrates very nicely the dynamics of the flow during the entire washing-out process. Figure 9.70a displays the depth-averaged concentration field when about 40% of the initial mass of scalar was washed out from the embayment. The coherent structures visualized by the concentration field in the experiment (Fig. 9.70b) are similar to those predicted by LES (Fig. 9.70a). The mass exchange is controlled by the engulfment of patches of high and, respectively, low concentration fluid by the large-scale structures present in the flow, as can be seen clearly from an animation showing the temporal evolution of the depth-averaged concentration field during the washing-out process (http://goo.gl/7YKwD).

Case 2 corresponds to the LES of flow hydrodynamics and mass exchange processes at a groyne field (Fig. 9.71) conducted by McCoy et al. (2008) and Constantinescu et al. (2009). These authors simulated one of the scaled-model test cases examined in the laboratory by Uijttewaal et al. (2001). The impermeable groynes were fully emerged and perpendicular to the channel bank. The embayments were shallow ($W = 7.5H$, $L = 10.75H$, with channel depth $H = 0.1$ m). The embayment bottom had a small slope toward the main channel. The embayment depth at the interface was $h_i = 0.6H$ and the average embayment depth was $h_m = 0.5H$. The CFD model contained only the first 6 embayments out of the 10 present in the scaled-model experiment. The incoming flow was fully turbulent and obtained from a precursor LES. Similar to Case 1, a free slip boundary condition was used at the lateral boundary opposite the groyne field. An outflow convective boundary condition was used at the exit section. The Reynolds number in the main channel of depth H was Re = 35,000. The mean velocity in the main channel was $U = 0.35$ m/s. The mesh contained about 4 million cells. The grid points were concentrated near the wall surfaces so that the use of wall functions could be avoided. The LES was conducted using the dynamic Smagorinsky SGS model. Similar to Case 1, a passive scalar with $C = 1$ was introduced instantaneously (at time $t = 0$) over the whole volume of one of the embayments (2 or 5) after the flow became statistically steady. At the inflow, C was set equal to 0. A unit value was assumed for the molecular Schmidt number.

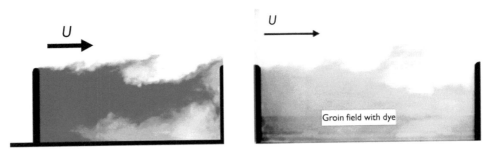

Figure 9.70 Flow and mass exchange between a channel and a groyne field: Visualization of the instantaneous depth-averaged scalar concentration field around an embayment (Case 1) when about 40% of the initial mass of scalar was washed out from the embayment. The left frame shows LES of Hinterberger et al. (2007) – picture by courtesy of C. Hinterberger. The right frame shows a snapshot of the scalar concentration field from dye experiments. The right frame is from Weitbrecht et al. (2008), reprinted with permission from ASCE.

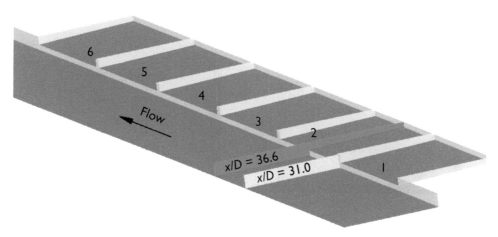

Figure 9.71 Flow and mass exchange between a channel and a groyne field: Computational domain for Case 2 with six shallow embayments. The cross section at $x/H = 36.6$ shows that the embayment bed is inclined toward the main channel and the flow depth is larger in the main channel. From McCoy et al., (2008), reprinted with permission from ASCE.

Consistent with the experiment and theory, a one-gyre circulation was observed in Figure 9.72 inside the embayments ($W/L = 0.7 > 0.5$). A smaller secondary counter-rotating gyre was present near the upstream corner of the embayment sidewall. Similar to the experiment, the size of this gyre increases with the rank of the embayment. The 2-D streamlines show that, on average, fluid from the channel enters the embayment close to the bottom and moves into the main channel close to the free surface. The fact that the transverse velocity is oriented toward the main channel close to the free surface is consistent with the measured mean flow pattern at the free surface in an experiment conducted by Weitbrecht et al. (2008) for embayments with $W/L = 1.12$, for which a one-gyre circulation was predicted (Fig. 9.73).

Comparisons of the instantaneous concentration field in Figure 9.74 (scalar initially introduced in embayment 2) show that most of the large-scale features in the concentration fields are qualitatively similar in a horizontal plane close to the free surface ($z = 0.95H$) and in a plane situated at $0.1H$ from the embayment bottom. A substantial amount of the scalar entrained into the interfacial mixing layer is then advected into the higher rank embayments before being eventually ejected back into the main channel. This is the main mechanism that increases the longitudinal dispersion of pollutant clouds in rivers with groynes. The maximum amount of scalar accumulated in embayments 3 to 6 is close to 20% of the initial mass of scalar introduced in embayment 2. At $t = 100H/U$ (Fig. 9.74a), the width of the region of relatively high-concentration fluid within the mixing layer decreases as one moves away from the free surface. The main reason is that, for the geometrical configuration of Case 2, most of the high-concentration fluid is ejected from embayment 2 into the mixing layer close to the free surface. Similar to Case 1, an important mechanism responsible for mass exchange is the random intrusion of highly-vortical eddies from the interfacial mixing layer region.

a)   b)

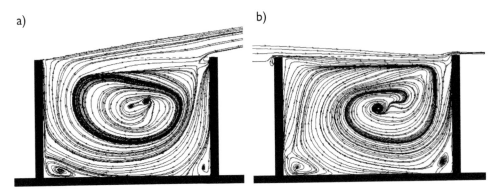

Figure 9.72 Flow and mass exchange between a channel and a groyne field: 2-D mean-flow streamline patterns around embayment 5 (Case 2). a) close to the free surface, $z/H = 0.95$; b) at a distance of $0.1H$ from the embayment bottom. From McCoy et al., (2008), reprinted with permission from ASCE.

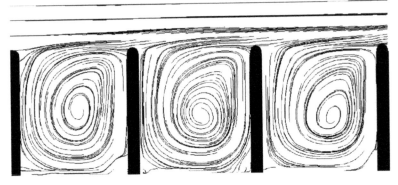

Figure 9.73 Flow and mass exchange between a channel and a groyne field: 2-D mean-flow streamline patterns visualized from an experiment conducted for a groyne field with $W/L = 1.12$. The flow depth in the embayment and in the main channel is the same. From Weitbrecht et al. (2008), reprinted with permission from ASCE.

Dead-zone theory models (e.g., see Uijttewall et al. 2001) are commonly used to characterize scalar washing-out processes. These models predict an exponential decay with time of the mean concentration inside the embayment. The characteristic time constant of the exponential decay, or mean residence time T_c, is defined as:

$$T_c = \frac{h_m W}{h_i U k} \tag{9.2}$$

where k is a non-dimensional exchange coefficient that characterizes the speed of the washing-out process. The time constant is determined by measuring the temporal decay of the mean concentration inside the embayment and by fitting an exponential decay curve which has a constant coefficient k. Based on results of several experiments,

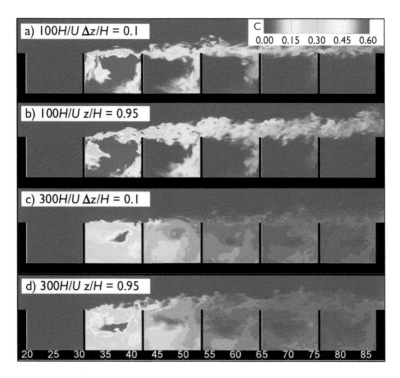

a) $100H/U$ $\Delta z/H = 0.1$

C
0.00   0.15   0.30   0.45   0.60

b) $100H/U$ $z/H = 0.95$

c) $300H/U$ $\Delta z/H = 0.1$

d) $300H/U$ $z/H = 0.95$

20   25   30   35   40   45   50   55   60   65   70   75   80   85

Figure 9.74 Flow and mass exchange between a channel and a groyne field: Instantaneous scalar concentration field (Case 2) in a plane close to the free surface ($z/H = 0.95$) and near the
embayment bed ($\Delta z/H = 0.1$) when about 20% ($t = 100H/U$) and 60% ($t = 300H/U$) of the
initial scalar mass was removed from embayment 2. From Constantinescu et al. (2009),
reproduced with kind permission from Springer Science + Business Media B.V.

Table 9.2 Comparison of the values of the mass exchange coefficient k obtained from experiments
and 3D LES simulations for mass exchange between a channel and a groyne field.

	R_D	k
Case 1		
3D LES (Hinterberger et al., 2007)	7.7	0.023
Experiment (Weitbrecht et al., 2008)	7.7	0.027
Case 2		
3D LES (Constantinescu et al., 2009)	4.42	0.018
Experiment (Uijttewaal et al., 2001)	4.42	0.016

Weitbrecht et al. (2008) proposed a linear variation of k with a non-dimensional
morphometric parameter called the embayment shape factor, R_D. This parameter is defined
by analogy to the non-dimensional hydraulic radius of the groyne field as $R_D = WL/$
$(W + L)h_m$. Results in Table 9.2 show that the LES predictions of k for both cases are in
good agreement with values determined from experimental measurements. The different
value of R_D explains the significantly different values of k in the two cases.

9.9 FLOW AND MASS EXCHANGE PROCESSES AROUND A CHANNEL-BOTTOM CAVITY

Pollutants, toxic substances or denser water may accumulate in bottom cavities and produce a stagnant pool which may be harmful to the environment. For instance, if the washing out of the pollutant or stagnant water inside cavity-like regions in rivers is too slow, fish that generally use these low velocity regions for shelter from predators or environmental stressors (e.g., pollution), feeding and reproduction purposes would have to avoid them. This is because the aeration, photosynthesis and other biological processes are adversely affected, making these regions uninhabitable for fish. Turbulent mixing phenomena between neutrally-buoyant or negatively-buoyant contaminants introduced inside the cavity and the turbulent overflow are of interest in many water-resources applications. They include flushing of saline water from bottom river cavities, flow over mining pits in rivers and accidental spills of contaminants or hazardous substances.

The mechanism of entrainment of the fluid situated initially inside the cavity is obviously dependent on the shape (e.g., rectangular, trapezoidal) and aspect ratio of the cavity, L/D, on the nature of the flow approaching the cavity (e.g., thickness and state of the boundary layer), the cavity Reynolds number (Re = UD/ν) and the initial Richardson number (Ri = $((\rho_1 - \rho_0)/\rho_0)gD/U^2$) which accounts for the density difference between the fluid inside the cavity and in the channel. Here, D is the cavity depth, L is the cavity length, U is the mean streamwise velocity in the flow over the cavity, and ρ_0 and ρ_1 are the initial fluid densities in the channel and inside the cavity, respectively. The x, y and z axes denote the streamwise, vertical and spanwise directions, respectively.

Based on 3-D LES of flow around a cavity with $L/D = 2$, Chang et al. (2006) have shown that the structural content of the flow in the cavity and the separated shear layer (SSL) forming over the cavity mouth is strongly dependent on the inflow conditions and, in particular, the state of the boundary layer. Further, the contaminant washing-out mechanism depends also on the way contaminant accumulates inside the cavity and how the flow over the cavity develops once the contaminant is introduced. One extreme case concerns the removal of contaminant or a negatively-buoyant fluid from bottom cavities by an impulsively started overflow (Armfield and Debler, 1993, Debler and Armfield, 1997). The other extreme case is when the channel flow is already established and the contaminant is introduced rapidly inside the cavity. For most real-life applications the conditions are somewhere in between.

In the following, well-resolved LES will be presented for the second case. In particular, the effect of the Richardson number (Ri = 0.0 and 0.2) on the removal of contaminant from a rectangular bottom cavity ($L/D = 2$, Re = 3,360, $U = 0.14$ m/s, $D = 0.024$ m) present in a long channel will be shown for the case of fully turbulent incoming flow. The channel Reynolds number is 20,470. More details are given in Chang et al. (2007a). The Boussinesq approximation was used in accounting for density and buoyancy effects. A unity value was assumed for the molecular Schmidt number, Sc. The contaminant was introduced instantaneously inside the cavity domain ($2D \times 1D \times 6D$) after the flow has become statistically steady. Details on the numerical method are available in Pierce and Moin (2004) and Chang et al. (2006). The dynamic subgrid-scale model was used to estimate the SGS viscosity and diffusivity.

The length of the computational domain in the spanwise direction was $6D$ and periodic boundary conditions were applied. A precursor LES of developed channel flow was used to obtain the mean and fluctuating velocity field at the inlet of the computational domain containing the cavity. The mesh contained close to 14 million cells. The first point off the solid surfaces was located at about 0.15 wall units. The cavity and channel walls were treated as no-slip smooth surfaces.

Figure 9.75 compares the mean streamline patterns predicted by LES for the non-buoyant case (Ri = 0.0) with those determined from experiment (Pereira and Sousa, 1994, 1995). The size and relative position of the two main recirculation zones inside the cavity are well predicted by LES. Profiles of the normal stress in the streamwise direction, $\overline{u'u'}$, and of the primary shear stress, $\overline{u'v'}$, predicted by LES at different streamwise locations are compared in Figure 9.76 with experiment. The peak values and the rates of decay of $\overline{u'u'}$ across the SSL are correctly predicted by LES at most streamwise locations. Near the cavity bottom, LES captures the secondary peak in $\overline{u'u'}$ at $x/D = 0.6$ and $x/D = 1.0$ associated the wall-jet-like current carrying large-scale

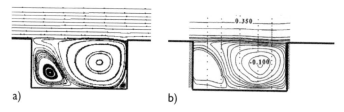

a)   b)

Figure 9.75 Flow and mass exchange around a channel-bottom cavity: Mean-flow streamlines in the case with a non-buoyant (Ri = 0.0) contaminant. a) LES; b) experiment of Pereira and Sousa (1994). From Chang et al. (2006). Reproduced with permission from the Cambridge University Press.

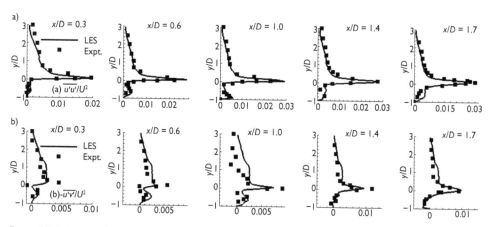

Figure 9.76 Flow and mass exchange around a channel-bottom cavity: Turbulent stresses in the case with a non-buoyant (Ri = 0) contaminant at different streamwise stations ($x/D = 0.0$ start of cavity; $x/D = 2.0$ end of cavity). Symbols correspond to experimental measurements of Pereira and Sousa (1994). a) $\overline{u'u'}/U$; b) $\overline{u'v'}/U$. From Chang et al. (2007a), reprinted with permission from ASCE.

turbulent eddies. The levels of $\overline{u'v'}$ inside the SSL are well predicted by LES, especially for $x/D > 1.4$. More validation for this case is discussed in Chang et al. (2006). Within the downstream part of the SSL, the most energetic frequency, f, expressed nondimensionally as a Strouhal number, $St = fD/U$, is equal to 0.38, which corresponds to the first mode predicted by theory (Rockwell, 1977) for cavities with $L/D = 2$.

None of the simulations with an incoming fully-developed turbulent flow contains large-scale spanwise vortices inside the SSL (Chang et al., 2006). Rather, as shown in Figures 9.77 and 9.78 that visualize the coherent structures using the Q criterion and vorticity magnitude contours, respectively, the downstream part of the SSL contains a wide array of highly 3-D eddies. Some of these eddies are partially or totally injected into the cavity close to the cavity trailing edge and form the jet-like flow (Lin and

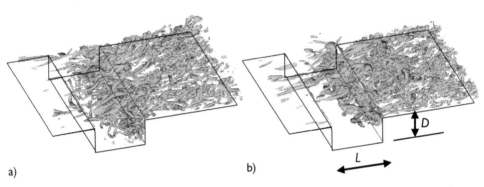

a)   b)

Figure 9.77 Flow and mass exchange around a channel-bottom cavity: Instantaneous coherent structures visualized using Q criterion in the case with: a) a non-buoyant (Ri = 0.0) contaminant; b) buoyant (Ri = 0.2) contaminant at $t = 85D/U$ after the start of the washing out process. From Chang et al. (2007a), reprinted with permission from ASCE.

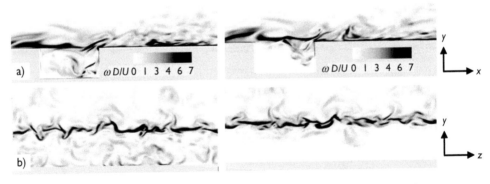

Figure 9.78 Flow and mass exchange around a channel-bottom cavity: Visualization of instantaneous flow structure using vorticity magnitude contours at $t = 85D/U$ after start of the washing out process: a) in a x-y section; b) in the $x/D = 1$ plane cutting through the middle of the cavity. The y and z axes correspond to the vertical and spanwise directions, respectively. Left: non-buoyant (Ri = 0.0) contaminant. Right: negatively-buoyant (Ri = 0.2) contaminant. From Chang et al. (2007a), reprinted with permission from ASCE.

Rockwell, 2001) that together with the eddies convected over the mouth of the cavity drive the mixing inside the cavity. In the simulation with Ri = 0.2, the large-scale eddies do not penetrate to large distances beneath the cavity-channel interface, at least not during the initial stages of the mixing process (e.g., see Fig. 9.77b corresponding to $t = 85D/U$ after the start of the washing-out process) when most of the heavier fluid (about 70% of the initial mass) is still located inside the cavity. This is because these eddies are damped as they try to penetrate a region of strong stable stratification. In fact, in the simulation with Ri = 0.2, most of the large-scale turbulence inside the cavity is contained within the roller vortex forming close to the trailing edge of the cavity.

The instantaneous concentration contours in Figure 9.79 visualize the way the high-concentration fluid/contaminant is removed from the cavity in the Ri = 0.0 case. At the start of the washing-out process (Fig. 9.79a), the mass transfer is mostly due to the engulfment of high-concentration fluid by the SSL eddies. As more and more eddies containing low-concentration fluid are deflected into the cavity at the trailing edge corner (Fig. 9.79b), mixing inside the cavity is enhanced. Most of the mixing inside the cavity occurs through these eddies, first in the region of the main downstream recirculation zone (Fig. 9.79c) and then in the upstream secondary recirculation zone (Fig. 9.79d).

In the buoyant case (Ri = 0.2) following the initial engulfment action due to the energetic SSL eddies, a stable roller vortex forms near the cavity trailing edge corner (Fig. 9.80a). This vortex has no equivalent in the Ri = 0.0 case. Furthermore, most of the mechanisms controlling the washing-out process are strongly affected by the stratification inside the cavity. For $t < 100D/U$, the tilting of the interface and the interfacial shear induced by the trailing edge vortex are the two dominating mechanisms. A third one, which plays a significant role for $t < 200D/U$, is the breaking of interfacial waves at the sharp density interface. Over the following phase ($100D/U < t < 200D/U$) of the washing-out process, the main mechanism is the scouring of the interface between low–and high-density fluid by the strong trailing-edge vortex situated on top of it (Fig. 9.80b). For $t > 200D/U$, the trailing vortex touches the cavity bottom and entrainment takes place mostly at the downstream side of the lower layer of higher density fluid (Figs. 9.80c and 9.80d).

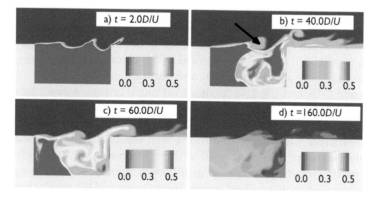

Figure 9.79 Flow and mass exchange around a channel-bottom cavity: Instantaneous contours showing evolution of contaminant concentration in an x-y plane from LES with a non-buoyant (Ri = 0.0) contaminant. From Chang et al. (2007a), reprinted with permission from ASCE.

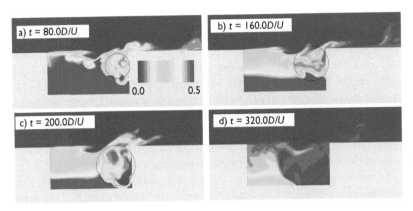

Figure 9.80 Flow and mass exchange around a channel-bottom cavity: Instantaneous contours showing evolution of contaminant concentration in an *x-y* plane from LES with a negatively buoyant (Ri = 0.2) contaminant. The circle shows approximately the region occupied by the roller vortex. From Chang et al. (2007a), reprinted with permission from ASCE.

As expected, the rate of decay of the mass of contaminant inside the cavity is smaller in the simulation with a negatively buoyant contaminant. In the case with Ri = 0.0, the decay is well approximated by an exponential function over the whole duration of the washing out process (Fig. 9.81). This confirms the validity of simple 1D dead-zone theory models (Uijttewaal et al., 2001) for neutrally-buoyant contaminants. The non-dimensional mass exchange coefficient appearing in Equation (9.2) in the case with Ri = 0.0 is $k = 0.013$. Except for very shallow cases in which the interfacial mixing layer and the flow within the embayments are significantly affected by bottom friction, the mechanisms responsible for the washing out of contaminant from a groyne field are similar to the ones observed in the non-buoyant cavity case discussed here. In the case with Ri = 0.2, analysis of the contaminant mass decay suggests the presence of three phases: a short initial one characterized by the scouring action of the SSL eddies near the interface, a second phase over which buoyancy effects are important and the mass decay is slow ($k \cong 0.006$ for $35D/U < t < 200D/U$), and a final one ($t > 200D/U$ after about 60% of the initial contaminant mass was removed from the cavity) over which buoyancy effects are no longer important and the decay rate is comparable ($k \cong 0.013$) to the one observed in the neutrally-buoyant case.

It is relevant here to mention that the neutrally-buoyant cavity case was also predicted using DES and URANS methods on much coarser meshes by Chang et al. (2007b). Both the SA and the SST versions of these methods were tested and the SA-DES was found to give the best overall predictions of the flow statistics and mass exchange coefficient when compared to the well-resolved LES discussed in this section. Calculations were also performed without fluctuations at the inflow (only the developed mean streamwise velocity profile was prescribed) and yielded inferior results. The effect of using resolved inflow fluctuations in DES was to break the large coherence of the vortices shed in the SSL present in the simulations with steady inflow conditions and to generate a wider range of 3-D eddies inside the cavity, similar to what is observed in well-resolved LES.

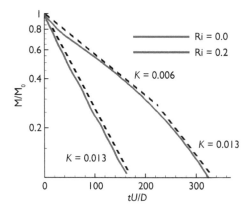

Figure 9.81 Flow and mass exchange around a channel-bottom cavity: Decay of mass of contaminant inside the cavity shown in log-linear scale for both a non-buoyant (Ri = 0) contaminant and a negatively-buoyant (Ri = 0.2) contaminant. The dashed lines identify the time intervals over which the decay of the mass is close to exponential. The corresponding value of the mass-exchange coefficient is also indicated in the figure. From Chang et al. (2007a), reprinted with permission from ASCE.

9.10 GRAVITY CURRENTS

Gravity currents are predominantly horizontal flows moving under the influence of gravity and are generated by density differences within a fluid or between two fluids. When gravity currents propagate over a loose bed, they can entrain, carry, and deposit large quantities of sediment over considerable distances from the entrainment location. A classical example is related to locking operations in an estuarine environment. Each time the facility is operated, a finite volume of saline water is released. Gravity currents are the main agents responsible for the transport of sediments from shallower to deeper regions in lakes and reservoirs (Meiburg and Kneller, 2010). Gravity currents can be generated by the instantaneous removal of a vertical lock gate separating two fluids at rest (Fig. 9.82a) and of different densities in a channel. The heavier lock fluid initially occupies the volume between the rear wall and the lock gate ($0 < x < x_0$, Fig. 9.82a). After the lock-gate is removed, a gravity current containing heavier lock fluid forms and propagates along the bottom wall.

In this section, we discuss LES of compositional currents where density differences are produced by a difference in a property of the fluid (e.g., temperature, salinity) and these differences are small enough for the Boussinesq approximation to be valid. The coupling between the Navier-Stokes equations and the transport equation for the concentration, $\overline{C}$, which is linearly related to the density ρ, occurs via the gravitational term in the equation for the vertical (y) component of the momentum. The filtered continuity and momentum equations and the advection-diffusion equation for $\overline{C}$ are made dimensionless using the channel depth, h, and the buoyancy velocity $u_b = \sqrt{g'h}$, where $g' = g(\overline{C}_{max} - \overline{C}_{min})/\overline{C}_{max}$, $\overline{C}_{max}$ and $\overline{C}_{min}$ are the maximum (lock fluid) and minimum (ambient fluid) initial concentrations in the domain. The dimensionless concentration is defined as $C = (\overline{C} - \overline{C}_{min})/(\overline{C}_{max} - \overline{C}_{min})$. The time scale is $t_0 = h/u_b$. The

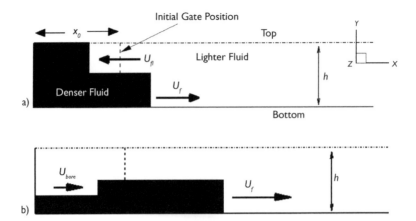

Figure 9.82 Sketch of a finite-volume lock-exchange flow in a channel. a) Lock release flow immediately after the gate is removed with associated front velocities of the heavier (U_f) and lighter (U_{fl}) currents; b) gravity current after the bore has formed. The y axis denotes the vertical direction. Courtesy of G. Constantinescu.

Reynolds number is Re = $u_b h/v$. The x, y and z axes denote the streamwise, vertical and spanwise directions, respectively.

The numerical solver is a finite-volume DNS/LES code (Pierce and Moin, 2001, 2004, Ooi et al., 2007). The conservative form of the Navier-Stokes equations is integrated on non-uniform Cartesian meshes. The algorithm is second order in space and time and discretely conserves energy, which allows obtaining solutions at high Reynolds numbers without artificial damping. All operators are discretized using central discretizations, except the convective term in the equation for $\overline{C}$, for which the QUICK scheme (Leonard, 1979) is used. The dynamic Smagorinsky model is used to calculate the SGS viscosity and eddy diffusivity (Pierce and Moin, 2001).

2-D and 3-D simulations were performed. The following discussion concentrates on the 3-D simulations in which the flow was assumed to be periodic in the spanwise direction. The length of the domain in the spanwise direction varied between h and $2h$. A zero normal gradient boundary condition was imposed for the concentration at the solid boundaries. C was initialized with a constant value of one in the region containing the lock fluid and a constant value of zero in the rest of the domain. The LES simulations were conducted on grids containing 30–200 million cells with 48–96 points in the spanwise direction. The mesh was stretched in the wall-normal direction of no-slip surfaces. The molecular Schmidt number was equal to 600, corresponding to saline diffusion in water. The initial height of the region containing lock fluid was equal to h.

9.10.1 Gravity currents propagating over a flat smooth bed

The advancement of the front position x_f of a gravity current with time t during the various phases of its propagation can be described as $x_f \sim t^\alpha$. Shallow water flow theory can be used to determine the value of α during each regime. A short time after the release of the lock gate (Fig. 9.82a) the front of the bottom-propagating current advancing over a horizontal smooth surface reaches a constant velocity, U_f, which is a function of the Reynolds number (Shin et al., 2004). This flow regime, in which x_f increases linearly with t (i.e. $\alpha = 1$), is called the slumping phase. As the backward propagating lighter current starts interacting with the rear wall, it reflects and forms a bore (Fig. 9.82b). The bore speed, U_{bore}, is nearly constant and slightly higher than U_f. For sufficiently high Reynolds numbers, once the bore catches the front, the heavier current transitions to the buoyancy-inertia self-similar phase in which U_f decays with time as $U_f \sim t^{-1/3}$. As $U_f = dx_f/dt$, this means that $x_f \sim t^{2/3}$. Once viscous effects become dominant, the current will transition to the viscous-buoyancy self-similar phase in which $U_f \sim t^{-4/5}$ (Huppert, 1982, Rottman & Simpson, 1983).

Next, we discuss some results of LES of lock-exchange bottom propagating currents based on the LES reported by Ooi et al. (2009). The other relevant case of lock exchange intrusion currents is discussed by Ooi et al. (2007) based also on LES results. The lock-exchange flow is dominated by two instabilities: the predominantly 2-D Kelvin Helmholtz (KH) instability at the interface between the heavier and the lighter (ambient) fluid, and the 3-D lobe-and-cleft instability at the front. These two instabilities are visualized in Figures 9.83 and 9.84 for the case of currents with a high volume of release (initial aspect ratio of the lock fluid $R = x_0/h >> 1$) in which the lock gate is positioned at the middle of the channel. No-slip conditions are applied at the top and bottom surfaces. In the low-Reynolds-number simulation (case LGR, Re = 3,150) the KH interfacial billows maintain their coherence and quasi 2-D character at large distances behind the fronts of the heavier bottom (forward)- and lighter top (backward)-propagating currents, in excellent agreement with the DNS of Härtel et al. (2000). By contrast, in the high Reynolds number simulation (case HGR, Re = 126,000), the flow is strongly turbulent behind the two fronts. The billows shed behind the head lose their coherence in the spanwise direction much more rapidly (over less than 2.5h behind the front of each current). As a result of the vortex stretching driven by the smaller highly 3-D eddies, the interfacial region is depleted of large scales at large distances behind the two fronts and is slightly tilted with respect to the horizontal (Fig. 9.83b).

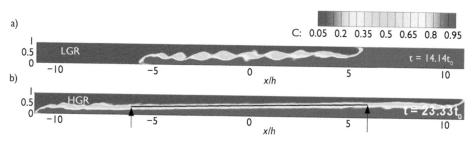

Figure 9.83 Gravity currents with a high volume of release propagating over a smooth flat bed: Visualization of the lock-exchange flow using spanwise-averaged concentration contours in a) case LGR (Re = 3,150); b) case HGR (Re = 126,000). The heavier fluid is initially positioned to the right of the lock gate ($x/h = 0$). From Ooi et al. (2009). Reproduced with permission from the Cambridge University Press.

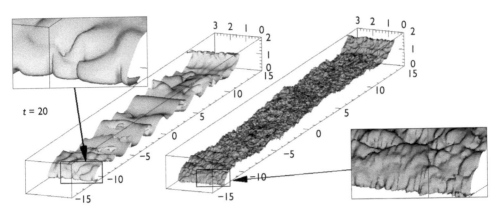

Figure 9.84 Gravity currents with a high volume of release propagating over a smooth flat bed: Visualization of gravity current interface using a concentration isosurface ($C = 0.5$) for cases LGR (left) and HGR (right). The insets show the development of the lobe and cleft structures at the front. Courtesy of G. Constantinescu.

Consistent with experiment and empirical formulas, the size of the lobes decreases with the increase of the Reynolds number (Fig. 9.84). The estimated average size of the lobes during the slumping phase in the LGR and HGR simulations is $0.5h$ and $0.13h$, respectively. The values inferred from experiments (Simpson, 1972) are $0.55h$ and $0.12h$, respectively. The non-dimensional front velocity, U_f/u_b, during the slumping phase in cases LGR (0.41) and HGR (0.46) is in excellent agreement with experiment. Using data from similar 3-D LES of currents with a high volume of release, Tokyay et al. (2012) found that at both low (Re $\cong 10^4$) and high (Re $\cong 10^5$–10^6) Reynolds numbers the turbulent kinetic energy in the tail of the current reaches a minimum at the location of the maximum streamwise velocity, consistent with the measurements of Kneller et al. (1999) and Stacey and Bowen (1988).

In the case of currents with a small initial volume of release (R $\cong 1$), the 3-D LES of Ooi et al. (2009) predicted $U_f/u_b \cong 0.45$–0.46 during the slumping phase for currents with Re = 7,000–20,000 and $0.67 < R < 1.5$, in good agreement with the corresponding experiments of Hacker et al. (1996). A simulation with Re = 10^6 and R = 1.78 (case CH) found $U_f/u_b \cong 0.485$, close to the value (0.48) measured experimentally by Keulegan (1957) for currents with Re > 10^5 and to the theoretical value (0.5) corresponding to an inviscid current. The comparison of the concentration distributions in Figures 9.85a (case CL-Re = 47,750) and 9.85d (case CH) close to the end of the slumping phase shows that the structure of the current is fairly insensitive to the Reynolds number for Re > 10^4. The 3-D LES of turbulent gravity currents with Re > 10^4 predicted that the bore overtakes the front (end of slumping phase) when $l = (x_f - x_0)/x_0 \cong 9$, in good agreement with experiment (Rottman and Simpson, 1983, Hacker et al., 1996). After the transition to the buoyancy-inertia phase is complete, the simulations predicted $U_f \sim t^\beta$ with the constant β getting very close to the theoretical value of $-1/3$ as the current approaches the inviscid limit (case CH). In fact, the temporal variation of x_f in cases CL and CH is very close.

Figure 9.85c shows the result of a 2-D simulation. Past the initial stages of the transition to the buoyancy-inertia phase, the mildly stratified tail region observed

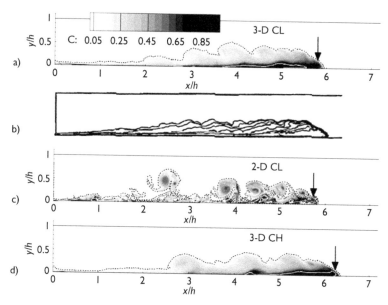

Figure 9.85 Gravity currents with a low volume of release propagating over a smooth flat bed: Concentration contours showing the structure of the gravity current at $t/t_0 = 12.4$ when the bore catches the front. a) spanwise-averaged contours, CL 3-D simulation; b) experimental results of Hacker et al. (1996) for case CL showing lines of constant dye concentration; c) CL 2-D simulation; d) spanwise-averaged contours, CH 3-D simulation. Position of the bore is indicated by an arrow. From Ooi et al. (2009). Reproduced with permission from the Cambridge University Press.

in the experiment (Fig. 9.85b) and 3-D simulation (Fig. 9.85a) is absent in the 2-D simulation. The interfacial KH vortices maintain their coherence (no vortex stretching by 3-D eddies is possible in a 2-D simulation) and the compact head region observed in experiment and 3-D LES is virtually absent in the 2-D simulation. This shows the limitation of 2-D eddy-resolving simulations.

9.10.2   Gravity currents propagating over a rough surface containing 2-D dunes or ribs or in a porous medium

In most practical applications, gravity currents propagate over loose surfaces containing large-scale natural bedforms (e.g., dunes, ripples) and/or topographic bumps and obstacles (Gonzalez-Juez et al., 2009, 2010). In others, like for stopping or slowing down powder-snow avalanches forming gravity currents, arrays of obstacles (e.g., ribs) are often used as protective measures on hilly terrains (Hopfinger, 1983). Another problem of great importance for environmental hazard mitigation is the dispersion of hyper-saline brines into rivers and coastal waters. Once the negatively-buoyant brine layer reaches the bed, it starts propagating away from the injection zone as a gravity current. The presence of a rough bed induces the accumulation of highly contaminated heavier brine fluid in the deeper regions of the bathymetry. In other cases, a gravity current advances through a layer of vegetation or through an array of flow retarding porous screens. Due to these obstacles, the structure of the current, its front velocity and its capacity to entrain sediment from the loose bed over which it propagates may change considerably with respect to the case of a current propagating over a smooth horizontal bed.

The additional drag force induced by large-scale roughness elements is a function of the height, shape and spacing of these elements. Tokyay et al. (2011a, 2011b, 2012) discusses in detail result of lock-exchange compositional currents with a high volume of release propagating over an array of 2-D dunes and square ribs based on results of 3-D LES. Here we discuss only four simulations conducted at Re = 48,000. In the base case, denoted LR-F, the current propagates over a flat bed (Fig. 9.86a). In the LR-D15 (Fig. 9.86b) and LR-R15 (Fig. 9.86c) simulations, the dunes and, respectively, the square ribs are of equal height ($D = 0.15h$) and equal wavelength ($\lambda = 3h$, $\lambda/D = 20$). To understand the effect of increasing the total drag force per unit streamwise length on the evolution of the current, an additional simulation, denoted LR-R15-HD with $D = 0.15h$ and $\lambda = h$ was performed (Fig. 9.87).

The obstacles and the backward-propagating hydraulic jumps forming each time the front overtakes an obstacle induce the formation of a layer of mixed fluid of variable height in cases LR-R15 and LR-D15 (Figs. 9.86b and 9.86c). The mixing at the front and over the tail is larger in case LR-R15 because ribs have a higher degree of bluntness compared to dunes of equal height. A strong jet-like flow forms downstream of the top/crest of each obstacle, a short time after the front begins to move away from that obstacle. In cases LR-R15 and LR-D15, the current reaches quickly a slumping phase and advances with $U_f \cong$ constant ($x_f \sim t$, i.e. $\alpha = 1$ see Fig. 9.88) until the front approaches the end of the channel. The front velocity during the slumping phase is smaller in the simulations with a rough bed and is a function of the degree of bluntness of the obstacles ($U_f/u_b \cong 0.34$, 0.4 and 0.45 for cases LR-R15, LR-D15 and LR-F, respectively).

In case LR-F, streaks of high and low u_* are present over most of the bottom wall surface between $x/h = 0$ and the front of the current (Fig. 9.86a), because in the strongly turbulent sections of the current the near-wall flow contains the usual coherent structures (e.g., streaks of low and high streamwise velocity) associated with a constant-density turbulent boundary layer. Streamwise velocity streaks are also present in cases LR-D15 (Fig. 9.86b) and LR-R15 (Fig. 9.86c), especially in the regions where the jet-like flow moves at a high speed parallel to the bottom surface. In the flat-bed case, the rapid initial growth of u_* at the current front is followed by a mild, nearly linear decay until $x/h = 0$ (Fig. 9.86d). Away from the head region, the distributions of u_* in cases LR-R15 and LR-D15 are strongly modulated by the obstacles and depend on the obstacle shape (Fig. 9.86d). u_* peaks where the jet-like flow impinges on the bottom wall.

The increase by a factor of three of the drag force per unit streamwise length in case LR-R15-HD (Fig. 9.87) compared to case LR-R15 explains why the current never reaches a slumping phase in the former case. Rather, after the front passes the first couple of ribs in the series, the current reaches a drag-dominated regime in which $x_f \sim t^\alpha$ with $\alpha = 0.72$ (Fig. 9.88). The exponent value (0.72) is close to the one (3/4) predicted by shallow water theory for high Reynolds number currents propagating in a porous medium during the quadratic drag-dominated regime (Fig. 9.90).

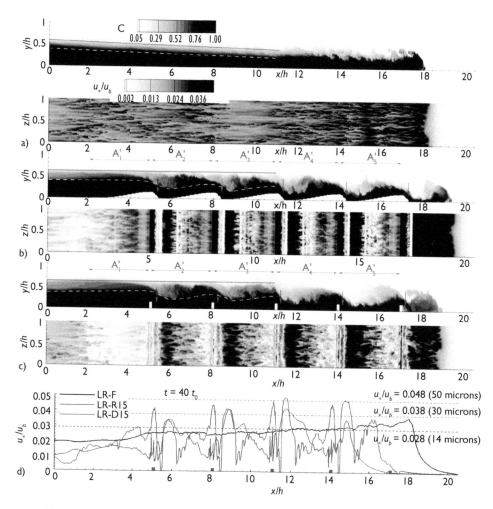

Figure 9.86 Gravity currents with a high volume of release propagating over a rough bed: Distributions of concentration in a vertical (x-y) section (top parts) and of bed friction velocity magnitude, u_*/u_b (bottom parts) when $x_f \cong 19h$. a) LR-F; b) LR-D15; c) LR-R15. The scale ratios x:y and x:z are 1:2 in the 2-D plots. The y and z axes correspond to the vertical and spanwise directions, respectively. The solid blue line shows the interface between the ambient fluid and the layer of mixed fluid. The dashed line shows the interface between the layer of mixed fluid and the bottom layer of heavier fluid. The arrows in frames b and c point toward the jet-like flow forming downstream of the top/crest of the obstacles. d) shows the distributions of the spanwise-averaged bed friction velocity magnitude in the three simulations at the time instants at which the currents are visualized in frames a, b and c. From Tokyay et al. (2012). Reproduced with permission from the Cambridge University Press.

Constantinescu (2011) discusses 3D LES of lock-exchange compositional currents with a high volume of release propagating into a porous medium of uniform porosity containing spanwise-oriented square cylinders. As the lock-exchange flow is close to anti-symmetric in all simulations, only the evolution of the bottom propagating

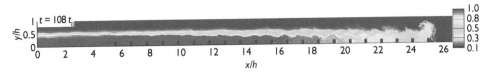

Figure 9.87 Gravity currents with a high volume of release propagating over a rough bed: Distributions of concentration in a vertical (x-y) section when $x_f \cong 25h$ in case LR-R15-HD. From Tokyay et al. (2011a). Reproduced with permission from the Cambridge University Press.

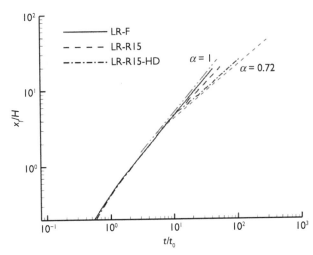

Figure 9.88 Gravity currents with a high volume of release propagating over a rough bed: Time variation of the non-dimensional position of the front, x_f/h, in the LR-F, LR-R15 and LR-R15-HD simulations plotted in log-log scale. The red and blue lines are $x_f \sim t^\alpha$ curves used to identify the slumping phase and the drag-dominated regime, respectively. The value of α is indicated next to each line. Courtesy of G. Constantinescu.

current is discussed. The main parameters describing the porous medium are the solid volume fraction, SVF, and the relative diameter of the cylinders, d/h. The first series of simulations were conducted with SVF = 12% and Re = 375 ($d/h = 0.05$) and 15,000 ($d/h = 0.035$), respectively. A second series of simulations were conducted with Re = 15,000 and SVF = 0%, 5% ($d/h = 0.035$) and 12% ($d/h = 0.035$), respectively. The channel length was 12h and the total number of cylinders was close to 600.

Simulation results show that for Re = 375, when the flow is basically laminar, the interface height defined by the C = 0.5 contour varies linearly with the streamwise distance from the lock gate up to the front position. This is in agreement with experiments conducted for low Reynolds number currents propagating into a vegetated canopy by Tanino et al., (2005). At high Reynolds numbers where the flow is turbulent, the concentration levels inside the head of the current are significantly lower in the case of a porous channel (Fig. 9.89b) compared to the non-porous case (Fig. 9.89a) due to the additional mixing induced by the interaction of the front with the cylinders, while the size of the eddies populating the interfacial region behind the head is smaller.

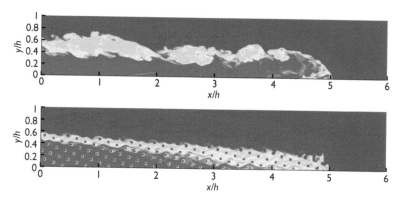

Figure 9.89 Gravity currents with a high volume of release propagating in a channel with and without a porous medium: Visualization of the structure of a bottom propagating current using concentration contours in a simulation with Re = 15,000, SVF = 0% (top) and Re = 15,000, SVF = 12% (bottom). Courtesy of G. Constantinescu.

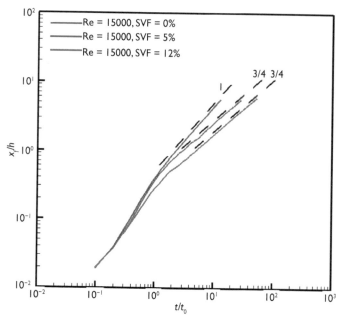

Figure 9.90 Gravity currents with a high volume of release propagating in a channel with and without a porous medium: Effect of solid volume fraction, SVF, on the temporal evolution of the front position, x_f/h. The dashed lines are $x_f \sim t^\alpha$ curves used to identify the slumping phase and the drag-dominated regime. The value of α is indicated next to each dashed line. Courtesy of G. Constantinescu.

As expected, results in Figure 9.90 show that the current does not leave the slumping phase (U_f = constant, or $\alpha = 1$) in the case with SVF = 0% until the end of the simulation. In the simulations conducted at low Reynolds numbers (Re = 375), the current reaches a regime where $x_f \sim t^{1/2}$ which is consistent with theory (Tanino et al., 2005) once the cylinder Reynolds number (Re_d) defined with U_f and d is smaller than one for the cylinders situated within the body of the current (linear drag regime, $C_D \sim 1/Re_d$). Regardless of the value of d/h, the simulations with Re = 15,000 predict $\alpha = 3/4$. Assuming a quadratic drag regime (C_D = constant), shallow water theory predicts $\alpha = 2/3$ (Hatcher et al., 2000), in fair agreement with LES (Fig. 9.90). LES predicts that the mean velocity in the bottom current at the lock gate position decays with $t^{-1/3}$, in excellent agreement with theory. This suggests that in the LES the current does not reach a self-similar regime. Most probably this is because shallow water theory neglects the effect of mixing, which is quite strong especially in the head region. More complex cases of lock-exchange currents propagating into vegetated canopies are discussed in Yuksel Ozan et al (2012).

9.11 ECO-HYDRAULICS: FLOW PAST AN ARRAY OF FRESHWATER MUSSELS

Freshwater mussels are bivalve mollusks that inhabit the substrates of rivers. Conservation and propagation of endemic freshwater mussels are crucial to maintain healthy river ecological systems (Cushing and Allan, 2001, Morales et al., 2006). A mussel individual has a shell separated into two symmetric valves surrounding the animal's body and mantle (Fig. 9.91). Mussels are partly burrowed in the sediment at the bottom of the stream. Freshwater mussels require substrates of sandy or gravelly material and flowing water to prevent siltation and enable transport of particulate organic material on which mussels feed. Nutrients are acquired by filtration of water by individual

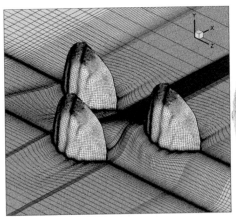

Figure 9.91 Flow past a cluster of mussels: Computational mesh on the surface of the mussels and on the surrounding region of the channel bottom. Also shown is a CAT-scanned image of a Washboard (*Megalonaias Nervosa*) mussel sample used to generate the mussel surface in the CFD model. From Constantinescu et al. (2013b), reproduced with permission from ASCE.

mussels. Hydrodynamic transport is essential for the survival of mussels. To investigate freshwater mussel habitats, several spatial scales have to be considered, including a basin scale, a habitat scale, and a mussel scale. The mussel scale is appropriate to investigate physical characteristics surrounding mussels. Flow structure affects the transport of nutrients, the settling and anchoring of the mussels on the bed and the stability of the mussels. In particular, local flow patterns around individual mussels affect nutrient availability to the mussel and contribute to the entrainment of nutrients into the benthic boundary layer. Numerical simulations could be used to: 1) obtain the flow fields around a mussel or a cluster of mussels with different mussel orientations; 2) study nutrient transport processes around them; and 3) estimate the drag force and the bed shear stress distributions which controls the local stability of the alluvial bed.

Constantinescu et al. (2013b) used LES to investigate the flow, the turbulence structure and the capability of the flow to induce scour around a cluster of three semi-buried Washboard mussels placed at the bottom of a flat-bed channel (Fig. 9.91). For the simplified arrangement considered, the flow field around the leading mussel is expected to resemble the one past an isolated mussel, while the ones past the trailing mussels will be partially affected by the wake and flow structures generated behind the most upstream mussel. Such wake-mussel interactions are characteristic for clusters with a more realistic arrangement of the mussels. Computerized Axial Tomography (CAT) was used to scan a 127-mm long, $W_m = 85.3$ mm wide, and $T_m = 53.5$ mm thick mussel sample. The high resolution CAT-scanned image retained all the relevant details of the real mussel surface (Fig. 9.91). The channel depth was $D = 5H_m = 0.318$ m, where H_m is the height of the exposed part of the mussel. The channel Reynolds number in the simulations was 35,000, corresponding to a mean channel velocity, U, of 0.11 m/s. The streamwise distance between the spanwise axes of the leading mussel and of the two trailing mussels was W_m. The spanwise distance between the stream-wise axes of the two trailing mussels was $1.5 W_m$. The 3-D mesh used contains around 4 million hexahedral elements arranged in an unstructured fashion. The dimensions of a typical computational cell inside the upstream part of the Separated Shear Layer (SSL) and the Horseshoe Vortex (HV) system were of the order of 10–25 wall units. The mesh close to the mussel shells and the channel bed was fine enough to avoid the use of wall functions.

A detailed description of the non-dissipative flow solver employed is given in Mahesh et al. (2004). The SGS viscosity was estimated with the dynamic Smagorinsky model. Data from Particle Image Velocimetry (PIV) were used to validate the LES method and to assess its predictive abilities with respect to those of RANS calculations with a k-ω SST model for the case in which the incoming flow is perpendicular to the shells of the mussels. The mesh (see Fig. 9.91) was the same in the RANS and LES calculations. Results from a precursor simulation corresponding to fully-developed turbulent channel flow were used to specify the inlet conditions in the domain containing the mussels. The free surface was treated as a shear-free rigid lid, which is justified as the channel Froude number was only 0.1 and the emerged part of the mussels occupied only a fifth of the channel depth. The channel bed and mussel surface were treated as smooth no-slip surfaces. The convective boundary condition was used at the outflow.

Experiment and LES show that the turbulent flow fields around the individual mussels are highly three-dimensional and dominated by energetic coherent structures that induce large-scale flow unsteadiness around the mussels and inside their wakes.

In particular, Figure 9.92 shows the presence of strongly-coherent vortex tubes in the upstream part of the SSL past each mussel. These highly energetic eddies control the mass exchange processes (e.g., suspended sediment, nutrients) between the recirculation flow region behind the mussels and the outer flow. Due to the shape of the mussels' edges, the axes of the vortex tubes have an arch-like shape rather than being oriented vertically, as found in the widely studied case of flow past vertical cylinders. However, formation and dynamics of these vortex tubes are similar to the ones forming in the SSLs around cylindrical bluff bodies which were discussed in sections 9.4 and 9.8.

Animations of instantaneous velocity fields obtained from PIV and LES show that the Horseshoe Vortex (HV) system in front of each mussel contains at all times one main horseshoe vortex which is subject to large-scale oscillations. This is expected from similar behavior found in flow around other surface-mounted bodies (e.g., cube, vertical plate). Additional secondary horseshoe vortices of variable coherence are shed from the region where the incoming boundary layer separates. Figures 9.93a and 9.93b compare the spanwise vorticity distributions and 2-D streamline patterns inferred from experiment and LES in a vertical plane cutting through the middle of one of the trailing mussels. Although the shape of the core of the main horseshoe vortex is slightly more elongated in LES, the circulation of the vortex predicted by LES is within 15% of the one estimated from the PIV measurements, which is a very good agreement. The orientation of the high-vorticity sheet corresponding to the SSL and the vorticity values inside it are similar in experiment and LES. By contrast, as

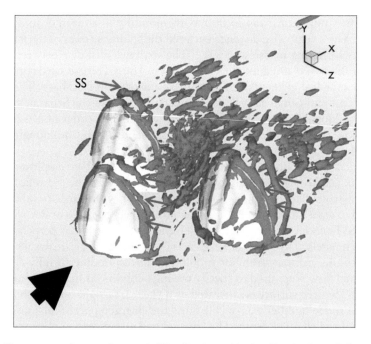

Figure 9.92 Flow past a cluster of mussels: Visualization with the Q criterion of the arch-shaped vortex tubes shed in the separated shear layers forming in between the overflow and the wake region behind the mussels. From Constantinescu et al. (2013b), reproduced with permission from ASCE.

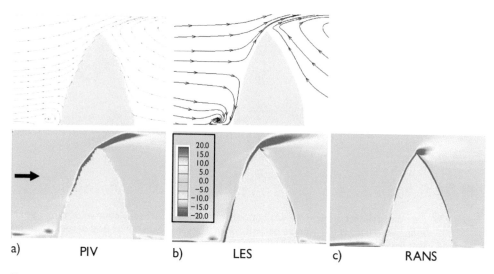

a)   PIV   b)   LES   c)   RANS

Figure 9.93 Flow past a cluster of mussels: Mean-flow 2-D streamline patterns (top) and spanwise vorticity, ω_z(1/s) (bottom), in a vertical plane cutting through the middle of one of the trailing mussels obtained from a) PIV; b) LES; c) RANS. Frames showing PIV and LES results are from Constantinescu et al. (2013b), reproduced with permission from ASCE. Frame showing RANS results is courtesy of G. Constantinescu.

shown by the vorticity contours in Figure 9.93c, RANS fails to correctly predict the horseshoe vortex ahead of the mussel and the vorticity in the SSL.

One of the most important issues from an environmental viewpoint is the stability of mussels, i.e. their capacity to resist their displacement from the channel bottom by the flow. Drag forces and moments induced by the water flow past the mussels and scouring around the mussels may decrease their stability and result in their dislocation, especially at flooding conditions. Given the good accuracy of LES in predicting the mean flow, LES predictions of the bed friction velocity, u_*, are expected to be sufficiently accurate to allow the determination of regions where scour will occur. High bed-friction velocity results in local scour and larger size of suspended sediment disturbs the substrate, thereby preventing juvenile mussel recruitment, and displaces the adult mussels. Too low bed-shear-stress levels result in deposition of fine silt and clay sediments, destroying mussel habitats (Strayer, 1999). As the flow contains large-scale coherent structures close to the bed, scour will occur not only in regions where the mean value of u_* is larger than the threshold value for entrainment of sediment of a given size, but also in regions where coherent structures can induce large instantaneous values of u_*. This is illustrated in Figure 9.94 by comparing LES predictions of mean bed friction velocity contours with contours at one time instant.

As expected for the case of a surface-mounted bluff body, the largest amplification of the mean u_* (Fig. 9.94aa) occurs in the regions of strong flow acceleration near the lateral edges of the mussels and between the trailing mussels, rather than beneath the HV system in front of the mussels. It can be seen in Figure 9.94b that the instantaneous u_* is strongly amplified beneath part of the region in which the main horseshoe

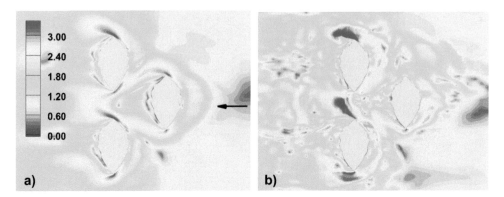

Figure 9.94 Flow past a cluster of mussels: Distribution of nondimensional bed-friction velocity magnitude, u_*/u_{*0}, where u_{*0} is the average bed-friction velocity in the incoming fully developed turbulent flow. a) LES, mean flow; b) LES, instantaneous flow. From Constantinescu et al. (2013b), reproduced with permission from ASCE.

vortex oscillates, particularly so for the right trailing mussel (with respect to the flow direction). Large-scale temporal fluctuations of u_* also occur within the jet-like flow region in between the two trailing mussels and beneath the HV region in front of the two trailing mussels.

The results presented clearly demonstrate that LES is a powerful tool for investigating flow and transport processes around clusters of mussels. Understanding the interactions among flow, sediment, nutrients and freshwater mussels, how flow affects the stability of the mussels, and the role of large-scale turbulence in controlling nutrient availability to individual mussels, is essential for a better comprehension of all aspects related to the life and behaviour of these aquatic organisms which play a very important role in river ecosystems and their habitat requirements.

9.12   FLOW IN A WATER PUMP INTAKE

Intake structures with one or more pumps are used to withdraw water from a river or a reservoir which is needed for irrigation, domestic and industrial use or in cooling systems of the electric and nuclear power generation plants. A water-intake structure includes an approach channel leading from the water source to a pump column which usually has an entrance structure called the suction or pump bell. It is commonly observed that water pumps are subjected to various operational problems such as impeller cavitation, uneven impeller loadings, noise and vibrations. All these phenomena result in a significant loss in the efficiency of the pumps (Melville et al., 1994). The flow inside a pump intake and in particular in the vicinity of the pump bell is generally characterized by the presence of free–surface and subsurface highly unsteady and, in some cases, intermittent vortices. These vortices form due to poor design or due to operating those structures outside their design range. The goal of the design/redesign process is not so much to eliminate these vortices, but rather to reduce their coherence (e.g., circulation) such that they cannot entrain air (in the case of non-pressurized intakes) and swirl inside the pump column is maintained below a threshold value so that the efficiency of the pumps is not significantly affected.

Until recent years, the only approach used to identify such problems and to find solutions was to conduct laboratory experiments using small-scale models. Most previous numerical investigations of flow and vortical structures at pump intakes used steady RANS simulations (e.g., see Constantinescu and Patel, 1998, 2000, Rajendran et al., 1999, Li et al., 2004). Such simulations have been shown to be subject to large errors in predicting the mean level of swirl inside the pump column, especially when strongly coherent unsteady vortices are present inside the intake. URANS simulations and LES-based techniques that have the capability to resolve the unsteady dynamics of the large-scale vortices offer a computationally more expensive, but potentially more successful, alternative. Such simulations were reported by Tokyay and Constantinescu (2006) for a pressurized pump intake and by Nakayama and Hisasue (2010) for an intake with a free surface.

In the present section, we focus on LES of flow in a pressurized pump intake of a nuclear power plant (Tokyay and Constantinescu, 2006) for which detailed validation data obtained using PIV are available from a scaled-model experimental study. Details of the numerical model used to perform the LES are given in Mahesh et al. (2004). The dynamic Smagorinsky model was used to evaluate the SGS viscosity. As shown in Figure 9.95a, the intake contains one pump column, two approaching channels and several flow training devices whose role is to decrease the coherence of the large

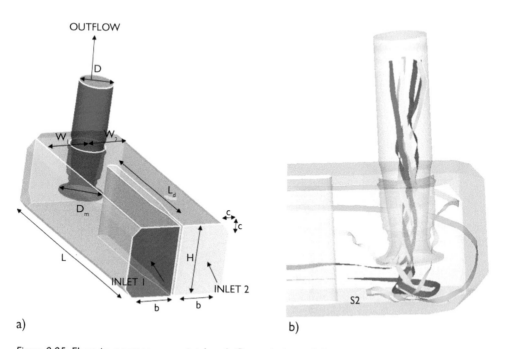

a)   b)

Figure 9.95 Flow in a water pump intake: a) General view of the pump intake geometry; b) 3-D streamtraces visualizing the swirling flow within the pump column. Courtesy of G. Constantinescu.

vortices. To better assess the predictive capabilities of LES, results of a steady RANS simulation using the SST model and of an URANS simulation using a Reynolds Stress Model (RSM) are also presented. The comparison allows estimating the performance of LES versus the one of typical simulations used by industry as a tool in the design/redesign of pump intake structures. Both LES and RANS simulations do not use wall functions.

The interior diameter of the downstream part of the pump column is chosen as reference length scale ($D = 130$ mm). The width and height of the two inlet channels are $b = 1.49D$ and $H = 1.9D$. The channel length is $L = 7.7D$ (Fig. 9.95a). The backwall of the intake is situated $0.6D$ from the centerline of the pump column. The pump column is placed symmetrically between the two sidewalls. The maximum diameter of the bell mouth is $D_m = 1.23_D$. For the experimental test case considered, the discharges through the two inlet sections are 0.905 and 0.0064 m³/s. This imbalance induces the formation of a strong floor-attached vortex. Figure 9.95b visualizes this vortex and the induced swirling motions within the pump column using 3-D streamtraces. The mean velocity inside the pump column at the location where its diameter is equal to D is taken as velocity scale, U. With this non-dimensionalization, the Reynolds number inside the pump column is 210,600. The mean velocities in the two inlet channels are $0.188U$ and $0.08U$, respectively. The Reynolds number defined with the average velocity of the two incoming channels and the channel height is about 45,000. The unstructured grid containing only hexahedral cells was generated using a paving technique that allows rapid clustering of grid cells in critical regions of the flow, while maintaining a good grid quality over the whole domain. Figure 9.96 shows several views of the LES mesh in a vertical section cutting through the symmetry axis of the pump column. The LES mesh contains over 4 million cells. The flexibility allowed by an unstructured mesh topology is evident in the very smooth transition (low stretching ratios) between coarse and fine regions needed to resolve the wall boundary layers. The mesh in the RANS calculations contains around 2.5 million cells. The first grid point off solid surfaces was situated

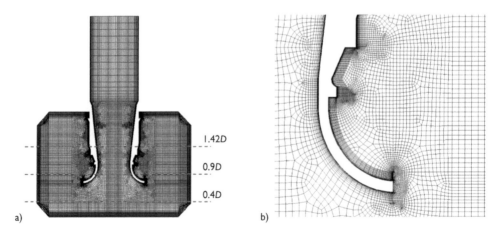

Figure 9.96 Flow in a water pump intake: View of the mesh. a) section through the center of the intake pipe parallel to the backwall; b) section showing meshing in the near-wall region of the pump column. From Tokyay and Constantinescu (2006), reprinted with permission from ASCE.

at 1–3 wall units in LES and RANS. The inflow conditions in LES and RANS were obtained from precursor LES/RANS simulations corresponding to fully developed turbulent flow in the two intake channels. The RANS simulations were conducted using the commercial code FLUENT, with the convective terms discretized using a second-order upwind scheme. The walls were treated as no-slip smooth surfaces.

Figure 9.97 compares the mean streamlines predicted by experiment, RANS and LES in several horizontal planes which are identified in Figure 9.96a. The imbalance between the discharges in the two incoming channels induces a strong horizontal shearing motion that takes the form of a bottom-attached vortex (Fig. 9.97a). This vortex is sucked into the pump column and is the main contributor to the swirl inside it. Because experimental data are not available inside the pump column, streamlines

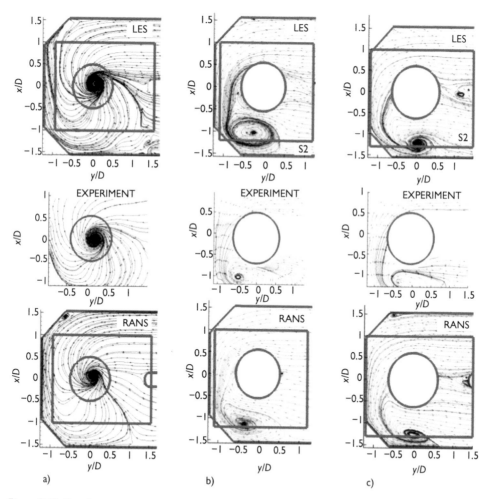

Figure 9.97 Flow in a water pump intake: 2D streamlines from LES, experiment and RANS in a horizontal plane situated a) at 0.4D away from channel bottom; b) at 0.9D away from channel bottom; and c) at 1.42D away from channel bottom. The positions of the three sections relative to the channel bottom are shown in Figure 9.96a. Courtesy of G. Constantinescu.

were not presented there in Figures 9.97b and 9.97c that visualize the flow in planes cutting through the pump column (Fig. 9.96a). In those planes, both RANS and LES predict the formation of another large vortex attached to the top wall, in between the pump column and the sidewall on the low-speed side of the intake. Overall, a good qualitative agreement with the mean streamlines inferred from the experimental data is observed for both simulations.

There are however important quantitative differences between the simulations in the distributions of the velocity and turbulent kinetic energy (tke) within the core of the bottom-attached vortex. Figure 9.98 compares these two quantities in a plane parallel to the sidewalls cutting through the centerline of the pump column. The core of large velocity extends very close to the bottom of the intake in the experiment and LES. By contrast, RANS predicts the core to penetrate only little beneath the mouth of the pump column. Though not shown, RSM-URANS predictions are close to SST-RANS. The differences are even more pronounced for the tke. In agreement with experiment, LES predicts a large amplification of the tke within the core of the bottom-attached vortex, starting very close to the channel bottom. RANS predicts a negligible level of amplification of the tke above the background turbulence levels in the surrounding flow. Rather, a large amplification of the tke is predicted inside of the pump column away of its centerline. This amplification is driven by the horizontal mean velocity gradient induced by the suction of the bottom-attached vortex into the pump column. While RSM-URANS fails to predict the amplification of the tke within the core of the bottom-attached vortex, the distribution of the tke within the pump column is qualitatively much closer to the LES than to the steady RANS prediction. These differences between the three simulations are illustrated in a more quantitative way in the bottom frames of Figure 9.98 presenting the velocity and tke profiles in a section situated mid-distance between the mouth of the pump column and the intake bottom. At this particular location, all predictions of the velocity show fairly noticeable disagreement with experiment. However, as opposed to URANS, LES predicts quite accurately the amplification of the turbulence within the core of the bottom-attached vortex. Analysis of the flow and turbulence statistics in other representative planes show that at most locations the LES statistics are in better agreement with experimental data compared not only with RANS (Tokyay and Constantinescu, 2006) but also to URANS.

Figure 9.99 compares the temporal evolution of the swirl (circulation associated with the streamwise vorticity over the whole cross section of the pump column) predicted by LES and RSM-URANS along with the (long-time) mean circulation in a section of the pump column situated 2D away from the intake bottom. RSM predicts significantly lower values of the mean circulation compared to LES, but both LES and RSM predict periods of time when the instantaneous circulation values are much larger (by couple of times) than the mean value. These variations are associated with large changes in the coherence of the bottom-attached vortex that is the main contributor to the swirl inside the pump column. Because LES resolves the flow structures down to grid scale, the time variation of the swirl is subject to both low-frequency oscillations resolved also by RSM and to higher frequency oscillations that are not captured by the URANS simulation. What is important for pump intake design is that the variance of the low-frequency fluctuations predicted by LES is significantly larger than the one predicted by RSM. This is significant, as even when the mean forces and moments are low, sharp large-scale variations of the forces and moments on the impeller blades can damage the pumps.

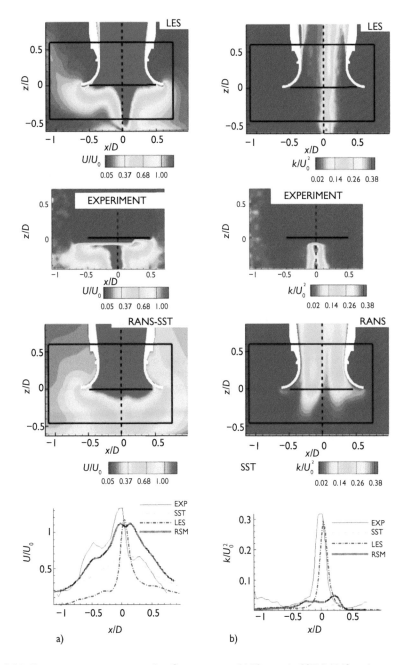

Figure 9.98 Flow in a water pump intake: Comparison of LES, steady SST RANS and unsteady RSM RANS results with experiment in a plane parallel to the sidewalls cutting through the centerline of the pump column. a) velocity magnitude; b) turbulent kinetic energy. The line plots at the bottom correspond to a cut at 0.35D from the bottom ($z/D = -0.3$). From Tokyay and Constantinescu (2006), reprinted with permission from ASCE.

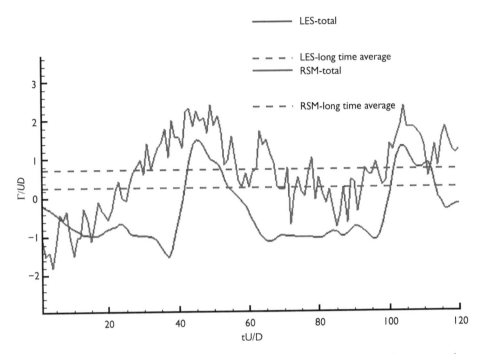

Figure 9.99 Flow in a water pump intake: Time history of circulation inside the pump column calculated in a horizontal plane situated close to the entrance. Comparison of URANS and LES predictions. The dashed lines indicate the time-averaged values. Courtesy of G. Constantinescu.

For instance, as pointed out by Johansson et al. (2005), the intermittent and direction-changing swirl inside the pump column may be more harmful to the pump efficiency than a steady one-directional swirl of similar magnitude. Obviously, a steady RANS simulation can give no information on the level of these large-scale fluctuations.

LES enabled a detailed investigation of the unsteady dynamics of the main coherent structures in a vorticity-dominated pump-intake flow and estimating variables of interest for design purposes. Though URANS was able to capture some of the large-scale unsteadiness in the flow and to give more accurate predictions compared to steady RANS, LES was the most successful method in predicting the mean flow and turbulence statistics compared to experiment. The RSM simulation predicted both a reduced level of the mean swirl and swirl fluctuations at different sections along the pump column compared to LES. Accurate prediction of the larger-scale swirl fluctuations inside the pump column is important for use of CFD as a design tool.

Appendix A

Introduction to tensor notation

This is a short introduction to Cartesian tensor notation, which is used extensively in this book. Cartesian tensor notation allows most equations to be written in a considerably more compact form than is possible with the conventional notation.

In Cartesian tensor notation, vector quantities are written by attaching an index to the symbol denoting the quantity, for example:

space vector: $x_i \equiv \{x_1, x_2, x_3\}$

velocity vector: $u_i \equiv \{u_1, u_2, u_3\}$

The three components of the vector in a Cartesian system (Figure A1) are obtained by setting the index (here i) equal to 1, 2 and 3, respectively.

A quantity with two indices (e.g. i and j) is called a tensor and has 9 components, which can be obtained by permutation of the 2 indices from 1 to 3.

$$a_{ij} \equiv \begin{Bmatrix} a_{11} & a_{12} & a_{13} \\ a_{21} & a_{22} & a_{23} \\ a_{31} & a_{32} & a_{33} \end{Bmatrix}$$

The subgrid stresses τ_{ij}^{SGS} appearing in the filtered momentum equation (2.10) are an example of such a tensor; here the first index denotes the surface ($\perp x_i$) on which the stress acts and the second index the direction of the stress. The (dyadic) product of two velocity vectors also yields a tensor expressing a momentum flux:

$$u_i u_j \equiv \begin{Bmatrix} u_1 u_1 & u_1 u_2 & u_1 u_3 \\ u_2 u_1 & u_2 u_2 & u_2 u_3 \\ u_3 u_1 & u_3 u_2 & u_3 u_3 \end{Bmatrix}$$

The tensors appearing in this book are all symmetric, that is $\tau_{ij} = \tau_{ji}$ and $u_i u_j = u_j u_i$ so that they have 6 different components.

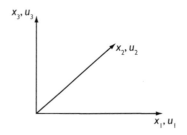

Figure A1 Cartesian coordinate system.

A particular tensor is the Kronecker delta δ_{ij} which has the components:

$$\delta_{ij} \equiv \begin{Bmatrix} 1 & 0 & 0 \\ 0 & 1 & 0 \\ 0 & 0 & 1 \end{Bmatrix}$$

so that $\delta_{ij} = 1$ for $i = j$ and $\delta_{ij} = 0$ for $i \neq j$.

One aspect of tensor notation remains to be explained which is particularly effective in making equations more compact. This is the summation convention which implies that whenever the same index is repeated in a single expression, the sum over all 3 directions has to be taken, thus

$$u_i u_i = \sum_{i=1}^{3} u_i u_i = u_1 u_1 + u_2 u_2 + u_3 u_3$$

The continuity equation (2.1) may be cited here as further example:

$$\frac{\partial u_i}{\partial x_i} = \sum_{i=1}^{3} \frac{\partial u_i}{\partial x_i} = \frac{\partial u_1}{\partial x_1} + \frac{\partial u_2}{\partial x_2} + \frac{\partial u_3}{\partial x_3} = 0$$

And the second term on the left hand side of the momentum equation (2.2) reads in full:

$$\frac{\partial u_i u_j}{\partial x_j} = \sum_{j=1}^{3} \frac{\partial u_i u_j}{\partial x_j} = \frac{\partial u_i u_1}{\partial x_1} + \frac{\partial u_i u_2}{\partial x_2} + \frac{\partial u_i u_3}{\partial x_3} = 0$$

References

Abe, K. (2005) A final LES/RANS approach using an anisotropy-resolving algebraic turbulence model. *International Journal of Heat and Fluid Flow*, 26, 204–222.

Aberle, J. & Nikora, V. (2006) Statistical properties of armored gravel bed surfaces. *Water Resources Research*, 42, W11414.

Adams, N.A., Hickel, S. & Franz, S. (2004) Implicit subgrid-scale modeling by adaptive deconvolution. *Journal of Computational Physics*, 200, 412–431.

Adrian, R.J. (2007) Hairpin vortex organization in wall turbulence. *Physics of Fluids*, 19(4), 041301.

Adrian, R.J., Christensen, K.T. & Liu, Z.C. (2000a) Analysis and interpretation of instantaneous turbulent velocity fields. *Experiments in Fluids*, 29(3), 275–290.

Adrian, R.J., Meinhart, C.D. & Tomkins, C.D. (2000b) Vortex organization in the outer region of the turbulent boundary layer. *Journal of Fluid Mechanics*, 422, 1–54.

Akselvoll, K. & Moin, P. (1993) Large eddy simulation of a backward-facing step flow. In: Rodi, W. & Martelli, F. (eds.) *Engineering Turbulence Modeling and Experiments 2*. Elsevier. pp. 303–313.

Anderson, W. & Meneveau, C. (2011) Dynamic roughness model for large-eddy simulation of turbulent flow over multiscale fractal-like rough surfaces. *Journal of Fluid Mechanics*, 679, 288–314.

Andren, A., Brown, A.R., Graf, J., Mason, P.J., Moeng C.H., Nieuwstadt, F.T.M. & Schumann, U. (1994) Large-eddy simulation of the neutrally stratified boundary layer: A comparison of four computer codes. *Quarterly Journal of the Royal Meteorological Society*, 120, 1457–1484.

Antonia, R.A. (1981) Conditional sampling in turbulence measurement. *Annual Review of Fluid Mechanics*, 13, 131–156.

Armenio, V., Piomelli, U. & Fiorotto, V. (1999) Effect of the subgrid scales on particle motion. *Physics of Fluids*, 11, 3030–3042.

Armfield, S.W. & Debler, W. (1993) Purging of density stabilized basins. *International Journal of Heat and Mass Transfer*, 36, 519–530.

Avdis, A., Lardeau, S. & Leschziner, M. (2009) Large eddy simulation of separated flow over a two-dimensional hump with and without control by means of a synthetic slot-jet. *Flow Turbulence and Combustion*, 83(3), 343–370.

Bagget, J.S. (1998) On the feasibility of merging LES with RANS for the near-wall region of attached turbulent flows. In: *Annual Research Briefs 1998*. Stanford, CA: Center for Turbulence Research, Stanford University.

Baggett, J.S., Jimenez, J. & Kravchenko, A.G. (1997) Resolution requirements in large-eddy simulation of shear flows. *CTR Annual Research Briefs*, 51–66.

Balaras, E., Benocci, C. & Piomelli, U. (1995) Finite-difference computations of high Reynolds-number flows using the dynamic subgrid-scale model. *Theoretical and Computational Fluid Dynamics*, 7, 207–216.

Balaras, E., Benocci, C. & Piomelli, U. (1996) Two layer approximate boundary conditions for large eddy simulations. *AIAA Journal*, 34, 1111–1119.

Bardina, J., Ferziger, J.H. & Reynolds, W.C. (1980) Improved subgrid scale models for large eddy simulations. AIAA Paper 80–1357.

Batten, P., Goldberg, U. & Chakravarthy, S. (2002) LNS – an approach toward embedded LES. AIAA Paper 2002–0427.

Batten, P., Goldberg, U. & Chakravarthy, S. (2004) Interfacing statistical turbulent closures with large eddy simulations. *AIAA Journal*, 42(3), 485–492.

Bennett, S.J. & Best, J.L. (1995). Mean flow and turbulence structure over fixed, two-dimensional dunes: Implications for sediment transport and bedform stability. *Sedimentology*, 42, 491–513.

Benocci, C., van Beeck, J.P.A.J. & Piomelli, U. (2008) Large-eddy simulation and related techniques: Theory and applications. VKI Lecture Series 2008–04.

Berselli, L.C., Illescu, T. & Layton, W. (2005) *Mathematics of Large-Eddy Simulation of Turbulent Flows*, Berlin: Springer.

Best, J.L. (2005) The fluid dynamics of river dunes: A review and some future research directions. *Journal of Geophysical Research*, 110, F04S02.

Bhaganagar, K., Kim, J. & Coleman, G. (2004) Effect of roughness on wall-bounded turbulence. *Flow Turbulence and Combustion*, 72, 463–492.

Blanckaert, K. (2010) Topographic steering, flow recirculation, velocity redistribution and bed topography in sharp meander bends. *Water Resources Research*, 46, W09506.

Blanckaert, K. & de Vriend, H.J. (2004) Secondary flow in sharp open-channel bends. *Journal of Fluid Mechanics*, 498, 353–380.

Bomminayuni, S. & Stoesser, T. (2011) Turbulence statistics of open-channel flow over a rough bed. *ASCE, Journal of Hydraulic Engineering*, 137(11), 1347–1358.

Boris, J.P., Grinstein, F.F., Oran, E.S. & Kolbe, R.L. (1992) New insights into large eddy simulation. *Fluid Dynamics Research*, 10, 199–208.

Breuer, M. (1998a) Large eddy simulation of the subcritical flow past a circular cylinder: Numerical and modeling aspects. *International Journal for Numerical Methods in Fluids*, 28, 1281–1302.

Breuer, M. (1998b) Numerical and modeling influences on large eddy simulations for the flow past a circular cylinder. *International Journal of Heat and Fluid Flow*, 19, 512–521.

Breuer, M. (2000) A challenging test case for large eddy simulation: High Reynolds number circular cylinder flow. *International Journal of Heat and Fluid Flow*, 21, 648–654.

Breuer, M. (2002) *Direkte Numerische Simulation und Large-Eddy Simulation Turbulenter Strömungen auf Hochleistungsrechnern*. Aachen: Shaker Verlag.

Breuer, M., Jovicic, N. & Mazaev, K. (2003) Comparison of DES, RANS and LES for separated flows around a flat plate at high incidence. *Int. J. Numer. Methods Fluids*, 41, 357–388.

Breuer, M., Jaffrezic, B. & Arora, K. (2007) Hybrid LES-RANS technique based on a one-equation near-wall model. *Theoretical and Computational Fluid Dynamics*, 22(3), 157–187.

Breuer, M. & Rodi, W. (1994) Large-eddy simulation of turbulent flow through a straight square duct and a 180 bend. In: Voke, P.R., Kleiser, L. & Chollet, J.P. (eds.) *Fluid Mechanics and its Applications*, Vol. 26. pp. 273–285.

Brooks, N.H. & Shukry, A. (1963) Discussion of "Boundary shear stress in curved trapezoidal channels" by A.T. Ippen and P.A. Drinker. *Journal of the Hydraulics Division, ASCE*, 89(HY3), 327–333.

Chan-Braun, C., García-Villalba, M. & Uhlmann, M. (2011) Force and torque acting on particles in a transitionally rough open channel flow. *Journal of Fluid Mechanics*, 684, 441–474.

Cabot, W.H. (1996) Near-wall models in large eddy simulations of flow behind a backward facing step. In: *Annual Research Briefs 1995*. Stanford, CA: Center for Turbulence Research, Stanford University. pp. 42–50.

Cabot, W.H. & Moin, P. (1999) Approximate wall boundary conditions in large-eddy simulation of high Reynolds number flows. *Flow, Turbulence, Combustion*, 63, 269–291.

Calhoun, R.J. & Street, R.L. (2001) Turbulent flow over a wavy surface: Neutral case. *Journal of Geophysical Research-Oceans*, 106, 9277–9293.

Cantero, M.I., Lee, J.R., Balachandar, S. & Garcia, M.H., (2007) On the front velocity of gravity currents. *Journal of Fluid Mechanics*, 586, 1–39.

Catalano, P., Wang, M., Iaccarino, G. & Moin, P. (2003). Numerical simulation of the flow around a circular cylinder at high Reynolds numbers. *International J. Heat and Fluid Flow*, 24, 463–469.

Cater, J.E. & Williams, J.J.R. (2008) Large eddy simulation of a long asymmetric compound channel. *Journal of Hydraulic Research*, 46(4), 445–453.

Cebeci, T. & Bradshaw, P. (1977) *Momentum Transfer in Boundary Layers*. New York, NY: McGraw-Hill.

Chang, K., Constantinescu, G. & Park, S.O. (2006) Analysis of the flow and mass transfer process for the incompressible flow past an open cavity with a laminar and a fully turbulent incoming boundary layer. *Journal of Fluid Mechanics*, 561, 113–145.

Chang, K., Constantinescu, G. & Park, S.O. (2007a) The purging of a neutrally buoyant or a dense miscible contaminant from a rectangular cavity. Part II: The case of an incoming fully turbulent overflow. *ASCE, Journal of Hydraulic Engineering*. 133(4), 373–385.

Chang, K., Constantinescu, G. & Park, S.O. (2007b) Assessment of predictive capabilities of detached eddy simulation to simulate flow and mass transport past open cavities. *ASME, Journal of Fluids Engineering*, 129(11), 1372–1383.

Chang, W.Y., Constantinescu, G., Tsai, W.F. & Lien, H.C. (2011) Coherent structures dynamics and sediment erosion mechanisms around an in-stream rectangular cylinder at low and moderate angles of attack. *Water Resources Research*, W12532. Available from: doi:10.1029/2011WR010586

Chapman, D.R. (1979) Computational aerodynamics development and outlook. *AIAA Journal*, 17, 1293–1313.

Chauvet, N., Deck, S. & Jacqin, L. (2007) Zonal detached eddy simulation of a controlled propulsive jet. *AIAA Journal*, 45, 2458–2473.

Choi, S.U. & Kang, H. (2004) Reynolds stress modeling of vegetated open channel flows. *Journal of Hydraulic Research*, 42(1), 3–11.

Chorin A.J. (1968) Numerical solution of the Navier–Stokes equations. *Mathematics of Computation*, 22, 745–762.

Chou, Y.J. & Fringer, O.B. (2010) A model for the simulation of coupled flow-bed form evolution in turbulent flows, *Journal of Geophysical Research*, 115, C10041. Available from: doi:10.1029/2010JC006103

Chow, F.K., Street, R.L., Xue, M. & Ferziger, J.H. (2005) Explicit filtering and reconstruction turbulence modeling for large-eddy simulation of neutral boundary layer flow. *Journal of the Atmospheric Sciences*, 62, 2058–2077.

Chu, V.H. & Babarutsi, S. (1988) Confinement and bed friction effects in shallow turbulent mixing layers. *Journal of Hydraulic Engineering*, 114, 1257–1274.

Colella, P. & Woodward, P. (1984) The piecewise parabolic method (PPM) for gas dynamics simulations. *Journal Computational Physics*, 54, 174–189.

Coleman, S. & Melville, B. (1996) Initiation of bed forms on a flat sand bed. *Journal of Hydraulic Engineering*, 122(6), 301–310.

Constantinescu, G. (2011) On the effect of drag on the propagation of compositional gravity currents, In: Rodi, W. & Uhlmann M. (eds.) *Environmental Fluid Mechanics*, IAHR Monograph, CRC Press. pp. 371–385.

Constantinescu, G., Chapelet, M.C. & Squires, K.D. (2003) Turbulence modeling applied to flow over a sphere. *AIAA Journal*, 41(9), 1733–1743.

Constantinescu, G., Koken, M. & Zeng, J. (2011a) The structure of turbulent flow in an open channel bend of strong curvature with deformed bed: Insight provided by an eddy resolving numerical simulation. *Water Resources Research*, 47, W05515.

Constantinescu, G., Miyawaki, S., Rhoads, B., Sukhodolov, A. & Kirkil, G. (2011b) Structure of turbulent flow at a river confluence with momentum and velocity ratios close to 1: Insights from an eddy-resolving numerical simulation. *Water Resources Research*, 47, W05507.

Constantinescu, G., Koken, M. & Zeng, J. (2012a) Flow and turbulence structure in shallow open channel bends. In: *CD-ROM Proceedings of 3rd IAHR International Symposium on Shallow Flows, 4–6 June 2012, Iowa City, IA, USA*.

Constantinescu, G., Miyawaki, S., Rhoads, B & Sukhodolov, A. (2012b) Numerical analysis of the effect of momentum ratio on the dynamics and sediment entrainment capacity of coherent flow structures at a stream confluence. *Journal of Geophysical Research – Earth Surface*, 117, F04028. Available from: doi:10.1029/2012JF002452

Constantinescu, G., Kashyap, S., Tokyay, T., Rennie, C.D. & Townsend, R.D. (2013a) Hydrodynamics processes and sediment erosion mechanisms in an open channel bend of strong curvature with deformed bathymetry. *Journal of Geophysical Research – Earth Surface*, Vol 118 1–17 doi:10.1002/jgrf.20042.

Constantinescu, G., Miyawaki, S. & Liao, Q. (2013b) Flow and turbulence structure past a cluster of freshwater mussels. *ASCE, Journal of Hydraulic Engineering*, Vol 139(4) 347–358.

Constantinescu, S.G. & Patel, V.C. (1998), Numerical model for simulation of pump intake flow and vortices. *ASCE, Journal of Hydraulic Engineering*, 124(2), 123–134.

Constantinescu, G.S. & Patel, V.C. (2000) Role of turbulence model in prediction of pump-bay vortices. *ASCE Journal of Hydraulic Engineering*, 126(5), 387–392.

Constantinescu, S.G., Sukhodolov, A. & McCoy, A. (2009) Mass exchange in a shallow channel flow with a series of groynes: LES study and comparison with laboratory and field experiments. *Environmental Fluid Mechanics*, 9(6), 587–615.

Constantinescu, G.S. & Squires, K.D. (2003) LES and DES investigations of turbulent flow over a sphere at Re = 10,000. *Flow, Turbulence and Combustion*, 70, 267–298.

Constantinescu, G. & Squires, K.D. (2004) Numerical investigation of the flow over a sphere in the subcritical and supercritical regimes, *Physics of Fluids*, 16(5), 1449–1467. Available from: doi:10.1063/1.1688325

Corino, E.R. & Brodkey, R.S. (1969) A visual investigation of wall region in turbulent flow. *Journal of Fluid Mechanics*, 37, 1–30.

Courant, R., Friedrichs, K. & Lewy, H. (1928) Über die partiellen Differenzengleichungen der mathematischen Physik (in German). Mathematische Annalen, 100(1), 32–74.

Cui, J. & Neary, V.S. (2002) Large-eddy simulation (LES) of fully developed flow through vegetation. In: *IAHR's 5th International Conference on Hydroinformatics, 1–5 July 2002, Cardiff, UK*.

Cui, J. & Neary, V.S. (2008) LES study of turbulent flows with submerged vegetation. *Journal of Hydraulic Research*, 46, 307–316.

Cui, J., Patel, V.C. & Lin, C.L. (2003) Prediction of turbulent flow over rough surfaces using a force field in large eddy simulation. *Journal of Fluids Engineering-Transactions of the ASME*, 125, 2–9.

Cushing, C.E. & Allan, J.D. (2001) *Streams: Their Ecology and Life*. San Diego, CA: Academic Press.

Dancey, C.L., Balakrishnan, M., Diplas, P. & Papanicolaou, A.N. (2000) The spatial inhomogeneity of turbulence above a fully rough, packed bed in open channel flow. *Experiments in Fluids*, 29, 402–410.

Dargahi, B. (1989) The turbulent flow field around a circular cylinder. *Experiments in Fluids*, 8, 1–12.

Davidson, L. (1998) Large eddy simulation: A dynamic one equation subgrid model for three dimensional recirculating flow. In: *Proceedings of the 11th Turbulent Shear Flows, 8–11 September 1997, Grenoble, France*, 3, 26.1–26.6.

Davidson, L. & Dahlstrom, S. (2005) Hybrid LES-RANS: An approach to make LES applicable at high Reynolds number. *International Journal of Computational Fluid Mechanics*, 19(6), 415–427.

Davidson, L. & Peng, S.H. (2003) Hybrid LES-RANS modelling: A one-equation SGS model combined with a k–w model for predicting recirculating flows. *International Journal of Numerical Methods in Fluids*, 43(9), 1003–1018.

Deardorff, J.W. (1974) Three dimensional numerical study of turbulence in an entraining mixed layer. *Boundary-Layer Meteorology*, 7, 199–226.

Debler, W. & Armfield, S.W. (1997) The purging of saline water from rectangular and trapezoidal cavities by an overflow of turbulent sweet water. *Journal of Hydraulic Research*, 1, 43–62.

Deck, S. (2005) Zonal detached eddy simulation of the flow around a high-lift configuration with deployed slat and flap. *AIAA Journal*, 43, 2372–2384.

Detert, M., Nikora, V. & Jirka, G.H. (2010) Synoptic velocity and pressure fields at the water-sediment interface of streambeds. *Journal of Fluid Mechanics*, 660, 55–86.

Devenport, W.J. & Simpson, R.L. (1990) Time-dependent and time-averaged turbulence structure near the nose of a wing–body junction. *Journal of Fluid Mechanics*, 210, 23–55.

Diurno, G.V., Balaras, E. & Piomelli, U. (2001) Wall models for LES of separated flows. In: Guerts, B. (ed.) *Modern Simulation Strategies for Turbulent Flows*. pp. 207–222. Philadelphia, PA: RT Edwards.

Druault, P.A., Lardeau, S., Bonnet, J.P., Coiffet, F., Delville, J., Lamballais, E., Largeau, J. & Perret, L. (2004) Generation of three-dimensional turbulent inflow conditions for large-eddy simulation. *AIAA Journal*, 43(3), 447–456.

Durbin, P.A. & Pettersson Reif, B.A. (2001) *Statistical Theory and Modeling for Turbulent Flows*. Chichester: John Wiley & Sons, Ltd.

Durbin, P. (1995) Separated flow computations with the k ε $\overline{v^2}$ model. *AIAA Journal*, 33, 659–664.

Dwyer, M.J., Patton, E.G. & Shaw, R.H. (1997) Turbulent kinetic energy budgets from a large-eddy simulation of airflow above and within a forest canopy. *Boundary-Layer Meteorology*, 84, pp. 23–43.

Egorov, Y. & Menter, F.R. (2008) Development and application of SST-SAS turbulence model in the DESIDER project. In: Peng, S.H. & Haase, W. (eds.) *Advances in Hybrid RANS-LES Modelling, Vol. 97 of Notes on Numerical Fluid Mechanics and Multidisciplinary Design*. Berlin: Springer. pp. 261–270.

Egorov, Y., Menter, F.R., Lechner, R. & Cokljat, A. (2010) The scale-adaptive simulation method for unsteady turbulent flow predictions. Part 2: Application to complex flows. *Flow, Turbulence and Combustion*, 85, 139–165.

Einstein, H.A. (1942) Formula for the transportation of bed load. *ASCE Transactions*, 107, 561–597.

Escauriaza, C., Sotiropoulos, F. (2011a) Reynolds number effects on the coherent dynamics of the turbulent horseshoe vortex system. *Flow Turbulence and Combustion*, 86(2), 231–262.

Escauriaza, C., Sotiropoulos, F. (2011b) Lagrangian model of bed-load transport in turbulent junction flows. *Journal of Fluid Mechanics*, 666, pp. 36–76.

Fadlun E.A., Verzicco R., Orlandi, P. & Mohd-Yusof J. (2000) Combined immersed-boundary finite-difference methods for three-dimensional complex flow simulations. *Journal of Computational Physics*, 61, 35–60.

Fan, T.C., Tian, M., Edwards, J.R., Hassan, H.A. & Baurle, R.A. (2004) Hybrid large eddy/ Reynolds averaged Navier–Stokes simulations of shock-separated flows. *Journal of Spacecraft Rockets*, 41(6), 897–906.

Ferguson, R.I., Parsons, D.R., Lane, S. & Hardy, R.J. (2003) Flow in meander bend with recirculation at the inner bank. *Water Resources Research*, 39(11), 1322. Available from: doi:10.1029/2003WR001965

Ferziger, J.H. (1996) Large eddy simulation, in *Simulation and Modeling of Turbulent Flows*, edited by T.B. Gatski, M.Y. Hussaini, & J.L. Lumley, Oxford Univ. Press, New York, 109–154.

Ferziger, J.H. & Peric, M. (2002) *Computational Methods for Fluid Dynamics*. 3rd edition. Berlin: Springer.

Fischer, H.B., List, E.J., Koh, R.C.Y., Imberger, J. & Brooks, N.H. (1979) *Mixing in Inland and Coastal Waters*. New York, NY: Academic Press.

Fischer-Antze, T., Stoesser, T., Bates, P.B. & Olsen, N.R. (2001) 3D numerical modelling of open-channel flow with submerged vegetation. *Journal of Hydraulic Research*, 39(3), 303–310.

Freund, J.B. (1997). Proposed inflow/outflow boundary condition for direct computation of aerodynamic sound. *AIAA J.* 35(4), 740–748.

Fröhlich, J. (2006) *Large-Eddy Simulation Turbulenter Strömungen*, Wiesbaden: Teubner Verlag.

Fröhlich, J. & von Terzi, D. (2008) Hybrid LES/RANS methods for simulation of turbulent flows. *Progress in Aerospace Sciences*, 44, 349–377.

Fröhlich, J, von Terzi, D., Severac, E. & Vaibar, R. (2010) Hybrid LES/RANS simulations: State of the art and perspectives. In: *Proceedings of the 8th International Symposium on Engineering turbulence Modelling and Measurements (ETMM8), 9–11 June 2010, Marseille, France.*

Fulgosi, M., Lakehal, D., Banerjee, S. & De Angelis, V. (2003) Direct numerical simulation of turbulence in a sheared air-water flow with a deformable interface. *Journal of Fluid Mechanics*, 482, 319–345.

Fureby, C. (1999) Large eddy simulation of rearward-facing step flow. *AIAA Journal*, 37, 1401–1410.

Fureby, C. & Grinstein, F.F. (1999) Monotonically integrated LES of free shear flows. *AIAA Journal*, 37, 544–550.

Fureby, C. & Grinstein, F.F. (2002) Large eddy simulation of high Reynolds number free and wall bounded flows. *Journal of Computational Physics*, 181, 68–97.

Galperin, B. & Orszag, S.A. (1993) *Large-Eddy Simulation of Complex Engineering and Geophysical Flows*. Cambridge: Cambridge University Press (republished March 2010).

Garbaruk, A., Shur, M. & Strelets, M. (2012) Turbulent flow past two-body configurations. UFR 2–12 document in the ERCOFTAC Knowledge Base Wiki. http://qnet-ercoftac.cfms.org.uk

Garcia-Villalba, M. & Fröhlich, J. (2006) LES of a free annular swirling jet – dependence of coherent structures on a pilot jet and the level of swirl. *International Journal of Heat and Fluid Flow*, 27, 911–923.

Garcia-Villalba, M., Li, N., Rodi, W. & Leschziner, M.A. (2009) Large-eddy simulation of separated flow over a three-dimensional axisymmetric hill. *Journal of Fluid Mechanics*, 627, 55–96.

Garnier, E., Mossi, M., Sagaut, P., Compte, P. & Delville, M. (1999) On the use of shock capturing schemes for large eddy simulation. *Journal of Computational Physics*, 153, 273–311.

Germano, M., Piomelli, U., Moin, P. & Cabot, W.H. (1991) A dynamic subgrid-scale eddy viscosity model. *Physics of Fluids A*, 3(7), 1760–1765.

Ghisalberti, M. & Nepf, H.M. (2002) Mixing layers and coherent structures in vegetated aquatic flows. *Journal of Geophysical Research*, 107, 3011.

Ghosal, S., Lund, T.S., Moin, P. & Akselvoll, K. (1995) A dynamic localization model for large-eddy simulation of turbulent flows. *Journal of Fluid Mechanics*, 286, 229–255.

Girimaji, S.S. (2006) Partially-averaged Navier–Stokes model for turbulence: A Reynolds-averaged Navier–Stokes to direct numerical simulation bridging method. *Journal of Applied Mechanics*, 73, 413–421.

Gonzalez-Juez, E., Meiburg, E. & Constantinescu, G. (2009) Gravity currents impinging on bottom mounted square cylinders: Flow fields and associated forces. *Journal of Fluid Mechanics*, 631, 65–102.

Gonzalez-Juez, E., Meiburg, E., Tokyay, T. & Constantinescu, G. (2010) Gravity current flow past a circular cylinder: Forces and wall shear stresses and implications for scour, *Journal of Fluid Mechanics*, 649, 69–102.

Grass, A.J. (1971) Structural features of turbulent flow over smooth and rough boundaries. *Journal of Fluid Mechanics*, 50, 233–255.

Grass, A.J., Stuart, R.J. & Mansour-Tehrani, M. (1991) Vortical structures and coherent motion in turbulent flow over smooth and rough boundaries. *Philosophical Transactions of the Royal Society of London, Series A*, 336(1640), pp. 33–65.

Grigoriadis, D.G.E., Balaras, E. & Dimas, A.A. (2009) Large-eddy simulations of unidirectional water flow over dunes, *Journal of Geophysical Research*, 114, F02022. Available from: doi:10.1029/2008JF001014

Grinstein, F.F. & Fureby, C. (2004) From canonical to complex flows: Recent progresses on monotonically integrated LES. *Computing in Science and Engineering*, 6, 34–69.

Grinstein, F.F., Fureby, C. & de Vore, C.R. (2005) On MILES based on flux-limiting algorithms. *International Journal of Numerical Methods in Fluids*, 47, 1043–1051.

Grinstein, F.F., Margolin, L.G. & Rider, W.G. (2007) *Implicit Large Eddy Simulation: Computing Turbulent Flow Dynamics*. Cambridge: Cambridge University Press.

Gritskevich, M., Garbaruk, A.V., Schutze, J. & Menter, F.R. (2012) Development of DDES and IDDES formulations for the k-w Shear Stress Transport model. *Flow, Turbulence, Combustion*, 88(3), 431–449.

Gullbrand, J. & Chow, F.K. (2003) The effect of numerical errors and turbulence models in large-eddy simulations of channel flow, with and without explicit filtering. *Journal of Fluid Mechanics*, 495, 323–341.

Guy, H., Simons, D. & Richardson, E. (1963) Summary of alluvial channel data from flume experiments. USGS Professional Paper, Vol. 96, pp. 1956–1961.

Hacker, J., Linden, P.F. & Dalziel, S.B. (1996) Mixing in lock-release gravity currents. *Dynamics of Atmospheres and Oceans*, 24, 183–195.

Hamba, F. (2003) A hybrid RANS/LES simulation of turbulent channel flow. *Theoretical and Computational Fluid Dynamics*, 16, 387–403.

Hanjalic, K. & Launder, B. (2011) *Modelling Turbulence in Engineering and the Environment – Second-Moment Routes to Closure*. Cambridge: Cambridge University Press.

Harlow, F.H. & Welch, J.E. (1965) Numerical calculation of time-dependent viscous incompressible flow of fluid with free surface. *Physics of Fluids*, 8, 2182.

Härtel, C., Meiburg, E. & Necker, F. (2000) Analysis and direct numerical simulation of the flow at a gravity-current head. Part 1: Flow topology and front speed for slip and no-slip boundaries. *Journal of Fluid Mechanics*, 418, 189–212.

Hatcher, L., Hogg, A.J. & Woods, A.W. (2000) The effects of drag on turbulent gravity currents. *Journal of Fluid Mechanics*, 416, 297–314.

Hey, R.D. (1978) Determinate hydraulic geometry of river channels. *ASCE, Journal of the Hydraulics Division*, 104, 869–885.

Hickel, S. & Adams, N.A. (2007) On implicit subgrid-scale modeling in wall bounded flows. *Physics of Fluids*, 19, 105106.

Hickel, S. & Adams, N.A. (2008) Implicit LES applied to zero-pressure-gradient and adverse-pressure-gradient boundary layer turbulence. *International Journal of Heat and Fluid Flow*, 29, 626–639.

Hickel, S., Adams, N.A. & Domaradzki, J.A. (2006) An adaptive local deconvolution method for implicit LES. *Journal of Computational Physics*, 213, 413–436.

Hickel, S., Adams, N.A. & Domaradzki, J.A. (2010) On the evolution of dissipation rate and resolved kinetic energy in ALDM simulations of the Taylor–Green flow. *Journal of Computational Physics*, 229, 2422–2433.

Hickel, S., Adams, N.A. & Mansour, N.N. (2007) Implicit subgrid scale modeling for large eddy simulation of passive scalar mixing. *Physics of Fluids*, 19, 095102.

Hickel, S., Kempe, T. & Adams, N.A. (2008) Implicit large eddy simulation applied to turbulent channel flow with periodic constriction. *Theoretical Computational Fluid Dynamics*, 22, 227–242.

Hinterberger, C., Fröhlich, J. & Rodi, W. (2007) Three-dimensional and depth-averaged large eddy simulations of some shallow water flows. *Journal of Hydraulic Engineering*, 133(8), 857–872.

Hinterberger, C., Fröhlich, J. & Rodi, W. (2008) 2D and 3D turbulent fluctuations in open channel flow with $Re\tau = 590$ studied by large eddy simulation. *Flow Turbulence and Combustion*, 80(2), 225–253.

Hinze, J.O. (1975) *Turbulence*. 2nd ed., New York: McGraw-Hill.

Hirsch, C. (1991) *Numerical Computation of Internal and External Flows*. West Sussex: John Wiley & Sons Ltd.

Hirsch, C. (2007) *Numerical Computation of Internal and External Flows*. 2nd edition. West Sussex: John Wiley & Sons Ltd.

Hirt, C.W. (1968) Heuristic stability theory for finite-difference equations. *Journal of Computational Physics*, 2, 339–355.

Hirt, C.W., Nichols, B.D. (1981) Volume of fluid (Vof) method for the dynamics of free boundaries. *Journal of Computational Physics*, 39, 201–225.

Hodges, B.R. & Street, R.L. (1999) On simulation of turbulent nonlinear free-surface flows. *Journal of Computational Physics*, 151, 425–457.

Holmes, P., Lumley, J.L. & Berkooz, G. (1996) *Turbulence, Coherent Structures, Dynamical Systems and Symmetry*. New York: Cambridge University Press.

Hopfinger, E.J. (1983) Snow avalanche motion and related phenomena. *Annual Review of Fluid Mechanics*, 15, 47–76.

Hoyas, S. & Jimenez, J. (2006) http://torroja.dmt.upm.es/~sergio/v_yx_summer.png related to the paper: Scaling of the velocity fluctuations in turbulent channels up to $Re\tau = 2003$. *Physics of Fluids*, 18, 011702.

Hunt, J.C.R., Wray, A.A. & Moin, P. (1988) *Eddies, Stream, and Convergence Zones in Turbulent Flows*. Center for Turbulence Research, Stanford University. Report number: CTR-S88.

Huppert, H. (1982) The propagation of two-dimensional and axisymmetric viscous gravity currents over a rigid horizontal surface. *Journal of Fluid Mechanics*, 121, 43–58.

Hyun, B.S., Balachandar, R., Yu, K. & Patel, V.C. (2003) Assessment of PIV to measure mean velocity and turbulence in water flow. *Experiments in Fluids*, 35, 262–267.

Iaccarino, G. & Verzicco, R. (2003) Immersed boundary technique for turbulent flow simulations. *Applied Mechanics Review*, 56, 331–347.

Israeli, M. & Orszag. S. (1981). Approximation of radiation boundary condition. *J. Comput. Phys.*, 41(1), 115–135.

Jeong, J., Hussain, F. (1995) On the identification of a vortex. *Journal of Fluid Mechanics*, 285, 69–94.

Jimenez, J. (2004) Turbulent flows over rough walls. *Annual Review of Fluid Mechanics*, 36, 173–196.

Johannesson, H. & Parker, G. (1989) Velocity redistribution in meandering rivers. *ASCE, Journal of Hydraulic Engineering*, 115(8), 1019–1039.

Johansson, A.E., Philip, S.S., White, D.K. & Fangbiao, L. (2005) Advancements in hydraulic modeling of cooling water intakes in power plants. In: *Proceedings of PWR 2005 ASME Power Conference, 5–7 April 2005, Chicago, IL, USA.*

John, V. (2004) *Large-Eddy Simulation of Turbulent Incompressible Flows: Analytical and Numerical Results for a Class of LES Models.* Berlin: Springer.

Kadota, A. & Nezu, I. (2002) Three dimensional structure of space-time correlation on coherent vortices generated behind dune crest. *Journal of Hydraulic Research*, 37(1), 59–80.

Kanda, M. & Hino, M. (1994) Organized structures in developing turbulent-flow within and above a plant canopy using a large-eddy simulation. *Boundary-Layer Meteorology*, 68, 237–257.

Kang, S. & Sotiropoulos, F. (2011) Flow phenomena and mechanisms in a field scale experimental meandering channel with a pool riffle sequence: Insights gained via numerical simulation. *Journal of Geophysical Research*, 116, F03011.

Kang, S., Lightbody, A., Hill, C. & Sotiropoulos, F. (2011) High resolution numerical simulation of turbulence in natural waterways. *Advances in Water Resources*, 34(1), 98–113.

Kara, S.J., Stoesser, T. & Sturm, T.W. (2012a) Turbulence statistics in compound open channels with deep and shallow overbank flows. *Journal of Hydraulic Research*, 50(5), 482–494.

Kara, S., Stoesser, T. & Sturm, T.W. (2012b) Effect of floodplain roughness on turbulence and mass and momentum exchange in a compound open channel, unpublished.

Kashyap, S., Constantinescu, G., Tokyay, T., Rennie, C. & Townsend, R. (2010) Simulation of flow around submerged groynes in a sharp bend using a 3D LES model. In: Dittrich, Koll, Aberle and Geisenhainer. (eds.) *International Conference on Fluvial Hydraulics, Proceedings of the River Flow 2010 Conference, Braunschweig, Germany.* Karlsruhe: Braunschweig: Bundesanstalt fur Wasserbau.

Kassinos, S., Langer, C., Iaccarino, G. & Moin, P. (2007) Complex effects in large-eddy simulations. In: *Lecture Notes in Computational Science and Engineering*, Vol. 65. Berlin: Springer.

Keating, A., Piomelli, U. (2006) A dynamic stochastic forcing method as a wall layer model for large eddy simulation. *Journal of Turbulence*, 7(12).

Keating, A., Pionelli, U. Balaras, E. & Kaltenbach, H.J. (2004) A priori and a posteriori tests of inflow conditions to large-eddy simulation. *Physics of Fluids*, 16(12), 4696–4712.

Keulegan, G.H. (1957) *An experimental study of the motion of saline water from locks into fresh water channels.* U.S. National Bureau Standards. Report number: 5168.

Kevlahan, N.K.R. (2007) Three-dimensional Floquet stability analysis of the wake in cylinder arrays. *Journal of Fluid Mechanics*, 592, 79–88.

Kim, J. (1987) Evolution of a vortical structure associated with the bursting event in a channel flow. In: Durst F., Launder B.E., Lumley J.L., Schmidt F.W. & Whitelaw J.H. (eds.), *Turbulent Shear Flows 5*, Berlin: Springer-Verlag.

Kim, J., Moin, P. & Moser, R. (1987) Turbulence statistics in fully-developed channel flow at low Reynolds-number. *Journal of Fluid Mechanics*, 177, 133–166.

Kim, S.J. & Stoesser, T. (2011) Closure modeling and direct simulation of vegetation drag in flow through emergent vegetation. *Water Resour. Res.*, 47(10), DOI: 10.1029/2011WR010561.

Kim, W.W. & Menon, S. (1995) A new dynamic one equation subgrid scale model for large eddy simulations. AIAA Paper 95–0356.

Kirkil, G. (2007) Numerical investigation of flow and sediment transport processes at cylindrical bridge piers at different stages of the scour process. Ph.D. Thesis, Department of Civil and Environmental Engineering, The University of Iowa.

Kirkil, G. & Constantinescu, S.G. (2008) A numerical study of shallow mixing layers between parallel streams. In: *CD-ROM Proceedings of 2nd IAHR International Symposium on Shallow Flows, December 2008, Hong Kong.*

Kirkil, G. & Constantinescu, S.G. (2009a) A numerical study of a shallow mixing layer developing over dunes. In: *XXXIIIrd International Association Hydraulic Research Congress, 9–14 August 2009, Vancouver, Canada.*

Kirkil, G. & Constantinescu, G. (2009b) Nature of flow and turbulence structure around an in-stream vertical plate in a shallow channel and the implications for sediment erosion. *Water Resources Research*, 45, W06412.

Kirkil, G. & Constantinescu, G. (2010) Flow and turbulence structure around an in-stream vertical plate with scour hole. *Water Resources Research*, 46, W1154.

Kirkil, G. & Constantinescu, G. (2012) The laminar necklace vortex system and wake structure in a shallow channel flow past a bottom-mounted circular cylinder. *Physics of Fluids*, 24, 073602. Available from: http://dx.doi.org/10.1063/1.4731291

Kirkil, G., Constantinescu, S.G. & Ettema, R. (2006) Investigation of the velocity and pressure fluctuations distributions inside the turbulent horseshoe vortex system around a circular bridge pier. In: *International Conference on Fluvial Hydraulics, River Flow 2006, September 2006, Lisbon, Portugal.*

Kirkil, G., Constantinescu, S.G. & Ettema, R. (2008) Coherent structures in the flow field around a circular cylinder with scour hole. *Journal of Hydraulic Engineering*, 134(5), 572–587.

Kirkil, G., Constantinescu, G. & Ettema, R. (2009) DES investigation of turbulence and sediment transport at a circular pier with scour hole. *Journal of Hydraulic Engineering*, 135(11), 888–901.

Kirkpatrick, M.P. & Armfield, S.W. (2005) Experimental and LES simulation results for the purging of salt water from a cavity by an overflow of fresh water. *International Journal of Heat and Mass Transfer*, 48, 341–351.

Klein, A., Sadiki, A. & Janicka, J. (2003) A digital filter based generation of inflow data for spatially developing direct numerical or large eddy simulations. *Journal of Computational Physics*, 186, 652–665.

Kline, S.J., Reynolds, W.C., Schraub, F.A. & Runstadl, P.W. (1967) Structure of turbulent boundary layers. *Journal of Fluid Mechanics*, 30, 741–773.

Kline, S.J. & Robinson, S.K. (1989) Quasi-coherent structures in the turbulent boundary layer: Part I. Status report on a community-wide summary of the data. In: *Near Wall Turbulence: 1988 Zaric Memorial Conference.* New York: Hemisphere.

Kneller, B., Bennett, S.J. & McCaffrey, W.D. (1999) Velocity structure, turbulence and fluid stresses in experimental gravity currents. *Journal of Geophysical Research Oceans*, 104, 5381–5391.

Koken, M. (2011) Comparison of coherent structures around isolated spur dikes at various angles. *IAHR, Journal of Hydraulic Research*, 49, 736–744.

Koken, M. & Constantinescu, G. (2008a) An investigation of the flow and scour mechanisms around isolated spur dikes in a shallow open channel. Part I. Conditions corresponding to the initiation of the erosion and deposition process. *Water Resources Research*, 44, W08406.

Koken, M. & Constantinescu, G. (2008b) An investigation of the flow and scour mechanisms around isolated spur dikes in a shallow open channel. Part II. Conditions corresponding to the final stages of the erosion and deposition process. *Water Resources Research*, 44, W08407.

Koken, M. & Constantinescu, G. (2009a) An investigation of the dynamics of coherent structures in a turbulent channel flow with a vertical sidewall obstruction. *Physics of Fluids*, 21, 085104. Available from: doi:10.1063/1.3207859

Koken, M. & Constantinescu, G. (2009b) A DES study of large-scale turbulence at a bridge abutment of realistic geometry with scour hole. In: *XXXIIIrd International Association Hydraulic Research Congress, 9–14 August 2009, Vancouver, Canada.*

Koken, M. & Constantinescu, G. (2011a) Flow and turbulence structure around a spur dike in a channel with a large scour hole. *Water Resources Research*, 47, W12511. Available from: doi:10.1029/2011WR010710

Koken, M. & Constantinescu, G. (2011b) Effect of Reynolds number and abutment geometry on the flow and turbulence structure around an abutment with a large scour hole. In: *XXXIVth International Association Hydraulic Research Congress, June 2011, Brisbane, Australia.*

Komori, S., Nagaosa, R., Murakami, Y., Chiba, S., Ishii, K. & Kuwahara, K. (1993) Direct numerical-simulation of 3-dimensional open-channel flow with zero-shear gas–liquid interface. *Physics of Fluids A – Fluid Dynamics*, 5, 115–125.

Kondo, K., Tsuchiya, M., Mochida, A. & Murakami, S. (2002) Generation of inflow turbulent boundary layer for LES computation. *Wind and Structures*, 5, 209–226.

Krajnovic, S. & Davidson, L. (2002) Large-eddy simulation of the flow around a bluff body. *AIAA Journal*, 40, 927–936.

Kriesner, B, Saric, S., Mehdizadeh, A., Jakirlic, S., Hanjalic, K., Tropea, C., Sternel, D., Gaub, F. & Schafer, M. (2007) Wall treatment in LES by RANS models: Method development and application to aerodynamic flows and swirl combustors. *ERCOFTAC Bulletin*, 72, 33–40.

Lam, K. & Banerjee, S. (1992) On the condition of streak formation in a bounded turbulent-flow. *Physics of Fluids A-Fluid Dynamics*, 4, 306–326.

Lam, K. & Lo, S.C. (1992) A visualization study of cross-flow around four cylinders in a square configuration. *Journal of Fluids and Structures*, 6, 109–131.

Lamb, M.P., Dietrich, W.E. & Venditti, J.G. (2008) Is the critical shields stress for incipient sediment motion dependent on channel-bed slope? *Journal of Geophysical Research*, 113, F02008. Available from: doi:10.1029/2007JF000831

Launder, B.E. & Sandham, N.D. (2002) *Closure Strategies for Turbulent and Transitional Flows*. Cambridge: Cambridge University Press.

Le, H., Moin, P. & Kim, J. (1997) Direct numerical simulation of turbulent flow over a backward-facing step. *Journal of Fluid Mechanics*, 330, 349–374.

Lee, S., Lele, S.K., Moin, P. (1992) Simulation of spatially evolving turbulence and the applicability of Taylor hypothesis in compressible flow. *Physics of Fluids A-Fluid Dynamics*, 4, 1521–1530.

Lefebvre de Plinvas-Salgues, H. (2004) *Implementation and evaluation of a method to generate turbulent inlet boundary conditions for a large eddy simulation*. Institute for Hydromechanics, University of Karlsruhe. Internal Report.

Leonard, A. (1974) Energy cascade in large eddy simulations of turbulent fluid flows. *Advances in Geophysics*, 18 A, 237–248.

Leonard, B.P. (1979) A stable and accurate convection modeling procedure based on quadratic upstream interpolation. *Computational Methods and Applications in Mechanical Engineering*, 19, 59–98.

Leonardi, S., Orlandi, P., Smalley, R.J., Djenidi, L. & Antonia, R.A. (2003) Direct numerical simulations of turbulent channel flow with transverse square bars on one wall. *Journal of Fluid Mechanics*, 491, 229–238.

Leschziner, M. & Rodi, W. (1979) Calculation of strongly curved open channel flow. *ASCE, Journal of Hydraulic Division*, 10, 1297–1313.

Lesieur, M., Métais, O. & Comte. P. (2005) *Large-Eddy Simulations of Turbulence*, Cambridge: Cambridge University Press.

Li, S.H., Lai, Y.G., Weber, L., Silva, J.M. & Patel, V.C. (2004) Validation of a three-dimensional numerical model for water-pump intakes. *Journal of Hydraulic Research*, 42(3), 282–292.

Lilly, D.K., (1992) A proposed modification of the Germano subgrid-scale closure method, *Physics of Fluids A*, 4, 633.

Lin, J.C. & Rockwell, D. (2001) Organized oscillations of initially turbulent flow past a cavity. *Journal of AIAA*, 39, 1139–1151.

Liu, D., Diplas, P., Fairbanks, J.D. & Hodges, C.C. (2008) An experimental study of flow through rigid vegetation. *Journal of Geophysical Research*, 113, F04015.

Liu, S. & Fu, S. (2003) Regimes of vortex shedding from an in-line oscillating circular cylinder in the uniform flow. *Acta Mechanica*, 19(2), 118–126.

Lopez, F. & Garcia, M.H., (2001) Mean flow and turbulence structure of open channel flow through non-emergent vegetation. *Journal of Hydraulic Engineering*, 127(5), 392–402.

Lumley, J.L. (1967) The structure of inhomogeneous turbulence. In: Yaglom, A.M. & Tatarski V.I. (eds.) *Atmospheric Turbulence and Wave Propagation*. pp. 166–78. Moscow: Nauka.

Lund, T., Wu, X. & Squires, K. (1998) Generation of turbulent inflow data for spatially-developing boundary layer simulations. *Journal of Computational Physics*, 140, 223–258.

Lyn, D.A. (1993) Turbulence measurements in open-channel flows over artificial bed forms. *Journal of Hydraulic. Engineering*, 119(3), 306–326.

Maddux, T.B., McLean, S.R. & Nelson, J.M. (2003a) Turbulent flow over 3D dunes. II: Fluid and bed stresses. *Journal of Geophysical Research*, 108, 17.

Maddux, T.B., Nelson, J.M. & McLean, S.R. (2003b) Turbulent flow over 3D dunes. I: Free surface and flow response. *Journal of Geophysical Research*, 108(F1), 20.

Mahesh, K., Constantinescu, G., Moin, P., (2004) A numerical method for large eddy simulation in complex geometries. *Journal of Computational Physics*, 197(1), 215–240.

Manes, C., Pokrajac, D., McEwan, I. & Nikora V. (2009) Turbulence structure of open channel flows over permeable and impermeable beds: A comparative study. *Physics of Fluids*, 21, 125109.

Margolin, L.G., Rider, W.J. & Grinstein, F.F. (2006) Modeling turbulent flows with implicit ILES. *Journal of Turbulence*, 7, 1–15.

Mathey, F. & Cokljat, D. (2005) Zonal multi-domain RANS/LES simulation of airflow over the Ahmed body. In: Rodi, W. & Mulas, M. (eds.) *Engineering Turbulence Modeling and Experiments*, Vol. 6. pp. 647–656.

Mathey F., Cokljat D., Bertoglio J.P. & Sergent E. (2003) Specification of inlet boundary condition using vortex method. In: Hanjalic K., Nagano Y. & Tummers M. (eds.) *Turbulence, Heat and Mass Transfer*, Vol. 4. West Redding, CT: Begell House, Inc.

Mathey, F., Cokljat, D., Bertoglio, J.P. & Sergent, E. (2006) Assessment of the vortex method for large eddy simulation inlet conditions. *Progress in Computational Fluid Dynamics*, 6, 58–67.

McCoy, A., Constantinescu, G. & Weber, L. (2006) Exchange processes in a channel with two emerged groynes. *Flow, Turbulence, Combustion*, 77, 97–126.

McCoy, A., Constantinescu, G. & Weber, L. (2007) A numerical investigation of coherent structures and mass exchange processes in a channel flow with two lateral submerged groynes. *Water Resources Research*, 43, W05445.

McCoy, A., Constantinescu, G. & Weber, L.J. (2008) Numerical investigation of flow hydrodynamics in a channel with a series of groynes. *Journal of Hydraulic Engineering*, 134(2), 157–172.

Meiburg E. & Kneller, B. (2010) Turbidity currents and their deposits. *Annual Review of Fluid Mechanics*, 42, 135–156.

Melaaen M.C. (1992) Calculation of fluid flows with staggered and non-staggered curvilinear non-orthogonal grids – the theory. *Numerical Heat Transfer B*, 21, 1–19.

Melville, B.W., Ettema, R. & Nakato, T., (1994) *Review of flow problems at water intake pump sumps*. Iowa Institute of Hydraulic Research, University of Iowa. EPRI Research Project RP3456–01 Final Report.

Mendoza, C. & Shen, H. (1990) Investigation of turbulent flow over dunes. *ASCE, Journal of Hydraulic Engineering*, 116(4), 459–477.

Meneveau, C., Lund, T.S. & Cabot, W. (1996) A Lagrangian dynamic subgrid-scale model of turbulence. *Journal of Fluid Mechanics*, 319, 353–385.

Menon S., Yeung P.K. & Kim W.W. (1996) Effect of subgrid models on the computed interscale energy transfer in isotropic turbulence. *Computers and Fluids*, 25(2), 165–180.

Menter, F.R. (1994) Eddy viscosity transport equations and their relation to the k–e model. NASA Technical Memorandum 108854, Ames Research Center, Moffett Field, CA, USA.

Menter, F.R. & Egorov, Y. (2005) A scale-adaptive simulation model using two-equation models. AIAA Paper 2005–1095.

Menter, F.R. & Egorov, Y. (2006) SAS turbulence modeling of technical flows. In: Lamballais, E., Friedrich, R., Guerts, B.J. & Metais, O. (eds.), *Direct and Large Eddy Simulation VI, 12–14 September 2005, Poitiers, France*. Berlin: Springer. pp. 687–694.

Menter, F.R., Egorov, Y. (2010) The scale adaptive simulation method for unsteady turbulent flow predictions. Part I: Theory and model descriptions. *Flow, Turbulence, Combustion*, 85, 113–138.

Menter, F.R., Egorov, Y. & Rusch, D. (2006) Steady and unsteady flow modeling using the model. In: Hanjalic, K., Nagano, Y. & Jakirlic, S. (eds.), *Turbulence, Heat and Mass Transfer 5*. New York: Begell House. pp. 403–405.

Menter F.R. & Kuntz, M. (2002). Adaptation of eddy-viscosity turbulence models to unsteady separated flow behind vehicles. In: McCallen, R., Browand, F. & Ross, J., (eds.), *The Aerodynamics of Heavy Vehicles: Trucks, Buses, and Trains*. New York: Springer. pp. 339–352.

Menter, F.R., Kuntz, M. & Bender, R. (2003) A scale-adaptive simulation model for turbulent flow predictions. AIAA Paper 2003–0767.

Menter, F.R., Garbaruk, A., Smirnov, P., Cokljat, D. & Mathey, F. (2009) Scala-adaptive simulation with artificial forcing. In: *3rd Symposium on Hybrid RANS-LES Methods*, Gdansk, Poland, June 2009.

Meyer, M., Hickel, S. & Adams, N.A. (2010) Assessment of implicit large-eddy simulation with a conservative immersed interface method for turbulent cylinder flow. *International Journal of Heat and Fluid Flow*, 31, 368–377.

Mierlo, M.C.L.M. & de Ruiter, J.C.C. (1988) Turbulence measurements above artificial dunes. Report Q789, Delft Hydraulics Laboratory, Delft, The Netherlands.

Miller, J.J.H. (1971) On the location of zeros of certain classes of polynomials with application to numerical analysis. *Journal of the Institute of Mathematics and its Applications*, 8, 397–406.

Miller, T.F. & Schmidt, F.W. (1988) Use of pressure-weighted interpolation method for the solution of the incompressible Navier–Stokes equations on a non-staggered grid system. *Numerical Heat Transfer*, 14, 213–233.

Mittal, R. & Iaccarino, G. (2005) Immersed boundary method. *Annual Review of Fluid Mechanics*, 37, 239–261.

Mittal, R. & Moin, P. (1997) Suitability of upwind biased schemes for large eddy simulation. *AIAA Journal*, 30(8), 1415–1417.

Mockett, C. & Thiele, F. (2007) Overview of detached eddy simulation for external and internal turbulent flow applications. *5th International Conference on Fluid Mechanics*, Shanghai, China.

Moeng, C.H. (1984) A large-eddy-simulation model for the study of planetary boundary-layer turbulence. *Journal of the Atmospheric Sciences*, 41, 2052–2062.

Moin P. & Moser R.D. (1989) Characteristic-eddy decomposition of turbulence in a channel. *Journal of Fluid Mechanics*, 155, 441–464.

Moin, P., Squires, K., Cabot, W. & Lee, S. (1991) A dynamic subgrid-scale model for compressible turbulence and scalar transport. *Physics of Fluids*, 3, 2746–2757.

Moncho-Esteve, I., Palau-Salvador, G., Shiono, K. & Muto, Y. (2010) Turbulent structures in the flow through compound meandering channels. In: Dittrich, Koll, Aberle and Geisenhainer (eds.) *Proceeding of River Flow 2010 Conference, Braunschweig, Germany*. Karlsruhe: Bundesanstalt fur Wasserbau.

Morales, Y., Weber, L.J., Mynett, A.E. & Newton, T.J. (2006) Mussel dynamics model: a tool for analysis of freshwater mussel communities. *Journal of Ecological Modeling*, 197, 448–460.

Moser, R.D., Kim, J. & Mansour, N.N. (1999) Direct numerical simulation of turbulent channel flow up to $Re_\tau = 590$. *Physics of Fluids*, 11, 943–946.

Muld T.W., Efraimsson, G. & Henningson D.S. (2012) Mode decomposition on surface mounted cube. *Flow, Turbulence and Combustion*, 88(3), 279–310.

Müller, A. & Gyr, A. (1986) On the vortex formation in the mixing layer behind dunes. *Journal of Hydraulic Research*, 24(5), 359–375.

Nadaoka, K. & Yagi, H. (1998) Shallow-water turbulence modeling and horizontal large-eddy computation of river flow. *ASCE, Journal of Hydraulic Engineering*, 124(5), 493–500.

Nakayama, A. & Hisasue, N. (2010) Large eddy simulation of vortex flow in intake channel of hydropower facility. *Journal of Hydraulic Research*, 48(4), 415–428.

Nakayama, A. & Sakio, K. (2002) *Simulation of flows over wavy rough boundaries*. Center for Turbulence Research. Annual Research Briefs.

Naot, D., Nezu, I. & Nakagawa, H. (1993) Calculation of compound channel flows. *ASCE, Journal of Hydraulic Engineering*, 119(12), 1418–1426.

Naot, D., Nezu, I. & Nakagawa, H. (1996) Hydrodynamic Behavior of partly vegetated open-channels. *ASCE, Journal of Hydraulic Engineering*, 122(11), 625–633.

Neary, V.S. (2000) Numerical model for open channel flow with vegetative resistance. In: *IAHR's 4th International Conference on Hydroinformatics, 23–27 July 2000, Iowa City, IA, USA*.

Nepf, H.M. & Vivoni, E.R., (2000) Flow structure in depth-limited, vegetated flow. *Journal of Geophysical Research* 105(C12), 28547–28557.

Nezu, I. & Nakagawa, H. (1993) *Turbulence in open-channel flows*, IAHR Monograph, Rotterdam: A.A. Balkema.

Nezu, I. & Rodi, W. (1986) Open-channel flow measurements with a laser Doppler anemometer. *Journal of Hydraulic Engineering*, 112, 335–355.

Nicoud, F. & Ducros, F. (1999) Subgrid-scale stress modelling based on the square of the velocity gradient tensor. *Flow, Turbulence and Combustion*, 62, 183–200.

Nikora, V., Goring, D., McEwan, I. & Griffiths, G. (2001) Spatially averaged open-channel flow over rough bed. *ASCE, Journal of Hydraulic Engineering*, 127, 123–133.

Nikora, V., McEwan, I., McLean, S., Coleman, S., Pokrajac, D. & Walters, R. (2007) Double-averaging concept for rough-bed open-channel and overland flows: Theoretical background. *ASCE, Journal of Hydraulic Engineering*, 133, 873–883.

Noh, W.F. & Woodward, P. (1976) SLIC (simple line interface calculation). In: van de Vooren A.I. & Zandbergen P.J. (eds.) *Proceedings of 5th International Conference of Fluid Dynamics*, Lecture Notes in Physics, Vol. 59, pp. 330–340.

Norberg, C. (2003) Fluctuating lift on a circular cylinder: Review and new measurements. *Journal of Fluids and Structures*, 17, 57–96.

Omidyeganeh, M. & Piomelli, U. (2011) Large-eddy simulation of two-dimensional dunes in a steady, unidirectional flow. *Journal of Turbulence*, 12(N42), 1–31.

Ong, L. & Wallace, J. (1996) The velocity field of the turbulent very near wake of a circular cylinder. *Experiments in Fluids*, 20, 441–453.

Ooi, S.K., Constantinescu, S.G. & Weber, L. (2007) A numerical study of intrusive compositional gravity currents. *Physics of Fluids*, 19, 076602. Available from: doi: 10.1063/1.2750672

Ooi, S.K., Constantinescu, S.G. & Weber, L. (2009) Numerical simulations of lock exchange compositional gravity currents. *Journal of Fluid Mechanics*, 635, 361–388.

Osher, S. & Sethian, J.A. (1988) Fronts propagating with curvature-dependent speed – algorithms based on Hamilton–Jacobi formulations. *Journal of Computational Physics*, 79, 12–49.

Paik, J., Escauriaza, C. & Sotiropoulos, F. (2007) On the bimodal dynamics of the turbulent horseshoe vortex system in a wing body junction. *Physics of Fluids*, 19, 045107.

Paik, J. & Sotiropoulos, F. (2005) Coherent structure dynamics upstream of a long rectangular block at the side of a large aspect ratio channel. *Physics of Fluids*, 17, 115104(11).

Pan, Y. & Banerjee, S. (1995) A numerical study of free-surface turbulence in channel flow. *Physics of Fluids*, 7, 1649–1664.

Pasche, E. & Rouve, G. (1985) Overbank flow with vegetatively roughened flood plains. *Journal of Hydraulic Engineering*, 111(9), 1262–1278.

Patel, V.C. (1998) Perspective: flow at high Reynolds number and over rough surfaces - achilles heel of CFD. *Journal of Fluids Engineering-Transactions of the ASME*, 120, 434–444.

Peller, N., Duc, A.L., Frédéric, T. & Manhart, M. (2006) High-order stable interpolations for immersed boundary methods. *International Journal of Numerical Methods Fluids*, 52(11), 1175–1193.

Pereira, J.C.F. & Sousa, J.M.M. (1994) Influence of impingement edge geometry on cavity flow oscillations. *AIAA Journal*, 32, 1737–1740.

Pereira, J.C.F. & Sousa, J.M.M. (1995) Experimental and numerical investigation of flow oscillations in a rectangular cavity. *Journal of Fluids Engineering*, 117, 68–73.

Pezzinga, G. (1994) Velocity distribution in compound channel flows by numerical modeling. *ASCE, Journal of Hydraulic Engineering*, ASCE, 120(10), 1176–1198.

Pierce, C.D. (2001) Progress variable approach for large eddy simulation of turbulence combustion. Ph.D. Thesis, Stanford University.

Pierce, C.D. & Moin, P. (2001) *Progress-variable approach for large-eddy simulation of turbulent combustion*. Stanford University. Mechanical Engineering Department Report TF–80.

Pierce, C.D. & Moin, P. (2004) Progress-variables approach for large-eddy simulation of non-premixed turbulent combustion. *Journal of Fluid Mechanics*, 504, 73–98.

Piomelli, U. (1999) Large-eddy simulation: Achievements and challenges. *Aerospace Sciences*, 35, 335–362.

Piomelli, U. & Balaras, E. (2002) Wall-layer model for large eddy simulations. *Annual Review of Fluid Mechanics*, 34, 349–374.

Piomelli, U., Balaras, E., Pasinato, H., Squires, K. & Spalart, P. (2003) The inner-outer layer interface in large eddy simulations with wall layer models. *International Journal of Heat and Fluid Flow*, 24, 538–549.

Piomelli, U., Ferziger, J., Moin, P. & Kim, J. (1989) New approximate boundary-conditions for large eddy simulations of wall-bounded flows. *Physics of Fluids A-Fluid Dynamics*, 1, 1061–1068.

Piomelli, U. & Liu, J. (1995) Large-eddy simulation of rotating channel flows using a localized dynamic model. *Physics of Fluids*, 7, 839–849.

Polatel, C. (2006) Large-scale roughness effect on free-surface and bulk flow characteristics in open-channel flows. Ph.D. Thesis, Iowa Institute of Hydraulic Research, The University of Iowa.

Pope, S. (2000) *Turbulent Flows*, Cambridge: Cambridge University Press.

Porte-Agel, F., Meneveau, C. & Parlange, M.B. (2000) A scale-dependent dynamic model for large-eddy simulation: Application to a neutral atmospheric boundary layer. *Journal of Fluid Mechanics*, 415, 261–284.

Rajaee, M., Karlsson, S. & Sirovich, L. (1995) On the streak spacing and vortex roll size in a turbulent channel flow. *Physics of Fluids*, 7(10), 2439–2443.

Rajendran, V., Constantinescu, S.G. & Patel, V.C. (1999) Experimental validation of a numerical model flow in pump intake bays. *ASCE, Journal of Hydraulic Engineering*, 125(11), 1119–1126.

Remmler, S. & Hickel, S. (2012) Direct and large-eddy simulation of stratified turbulence. *International Journal of Heat* and Fluid Flow, 35, 13–24.

Reynolds, W.C. (1990) The potential and limitations of direct and large-eddy simulations. In: Lumley J.L. (ed.) *Whither Turbulence? Turbulence at the Crossroads*, Lecture Notes in Physics. Berlin: Springer.

Rhie, C.M. & Chow, W.L. (1983) A numerical study of the turbulent flow past an isolated airflow with trailing edge separation. *AIAA Journal*, 21, 1525–1532.

Rhoads, B. & Sukhodolov, A. (2001) Field investigation of three-dimensional flow structure at stream confluences: Part I. Thermal mixing and time-averaged velocities. *Water Resources Research*, 37(9), 2393–2410.

Rider, W.J. & Margolin, L.G. (2003) From numerical analysis to implicit subgrid turbulence modelling. In: *Proceedings of the 16th AIAA CFD Conference, 23–26 June 2003, Orlando, FL, USA*. AIAA paper 2003–4101.

Robinson, S. (1991) Coherent motions in the turbulent boundary-layer. *Annual Review of Fluid Mechanics*, 23, 601–639.

Rockwell, D. (1977) Prediction of oscillation frequencies for unstable flow past cavities. *Journal of Fluids Engineering*, 99, 294–300.

Rodi, W. (1993) *Turbulence Models and their Application in Hydraulics*. 3rd edition. IAHR Monograph, Rotterdam: A.A. Balkema.

Rodi, W., Ferziger, J.H., Breuer, M. & Pourquie, M. (1997) Status of large eddy simulation: Results of a workshop. *Journal of Fluids Engineering-Transactions of the ASME*, 119, 248–262.

Rollet-Miet, P., Laurence, D. & Ferziger, J., (1999) LES and RANS of turbulent flow in tube bundles. *International Journal of Heat Fluid Flow*, 20, 241–254.

Rotta, J.C. (1972) *Turbulente Stromungen*. Stuttgart: Teubner.

Rottman, J.W. & Simpson, J.E. (1983) Gravity currents produced by instantaneous releases of a heavy fluid in a rectangular channel. *Journal of Fluid Mechanics*, 135, 95–110.

Rummel, A.C., Socolofsky, S.A., von Carmer, C.F. & Jirka, G.H. (2005) Enhanced diffusion from a continuous point source in shallow free-surface flow with grid turbulence. *Physics of Fluids*, 17, 075105.

Sagaut, P. (2006) *Large-Eddy Simulation for Incompressible Flows – An Introduction*. 3rd edition. Berlin: Springer.

Saito, N., Pullin, D.I. & Inoue, M. (2012) Large-eddy simulation of smooth-wall, transitional and fully rough-wall channel flow. Physics of Fluids, 24(7), 075103.

Salvetti, M.V. & Banerjee, S. (1995) *A priori* tests of a new dynamic subgrid-scale model for finite-difference large-eddy simulations. *Physics of Fluids*, 7, 2831–2847.

Salvetti, M.V., Zang, Y., Street, R.L. & Banerjee, S. (1997) Large-eddy simulation of free-surface decaying turbulence with dynamic subgrid-scale models. *Physics of Fluids*, 9, 2405–2420.

Sarghini, F., Piomelli, U. & Balaras, E. (1999) Scale-similar models for large-eddy simulations. *Physics of Fluids*, 11, 1596–1607.

Schluter, J.U., Pitsch, H. & Moin, P. (2004) LES inflow conditions for coupling with Reynolds averaged flow solvers. *AIAA Journal*, 42, 478–484.

Schluter, J.U., Wu, X., Kim, S., Shankaran, S., Alonso, J.J. & Pitsch, H. (2005) A framework for coupling Reynolds-averaged with large eddy simulations for gas turbine applications. *Journal of Fluids Engineering*, 127, 806–815.

Schumann, U. (1975) Subgrid scale model for finite-difference simulations of turbulent flows in plane channels and annuli. *Journal of Computational Physics*, 18, 376–404.

Schumann, U. (1993) Direct and large-eddy simulation of turbulence – summary of the state of the art 1993. In: *Introduction to Turbulence Modelling II*, VKI-Lecture Series number 1993–02. Brussels: Von Karman Institute for Fluid Dynamics.

Scotti, A. (2006) Direct numerical simulation of turbulent channel flows with boundary roughened with virtual sandpaper. *Physics of Fluids*, 18, 031701.

Shaw, R.H. & Schumann, U. (1992) Large-eddy simulation of turbulent-flow above and within a forest. *Boundary-Layer Meteorology*, 61, 47–64.

Shi, J., Thomas, T.G. & Williams, J.J.R. (2000) Free-surface effects in open channel flow at moderate froude and Reynold's numbers. *Journal of Hydraulic Research*, 38, 465–474.

Shimizu, Y. & Tsujimoto, T. (1994) Numerical analysis of turbulent open-channel flow over vegetation layer using a k–e turbulence model. *Journal of Hydroscience and Hydraulic Engineering*, 11(2), 57–67.

Shin, J., Dalziel, S., Linden, P.F. (2004) Gravity currents produced by lock exchange. *Journal of Fluid Mechanics*, 521, 1–34.

Shur, M.L., Spalart, P.R., Strelets, M.K. & Travin, A.K. (1999) Detached-eddy simulation of an aerofoil at high angle of attack. In: *Paper Presented at the Fourth International Symposium on Engineering Turbulence Modeling and Measurements, Corsica, France*.

Shur, M.L., Spalart, P.R., Strelets, M. & Travin, A. (2008) A hybrid RANS-LES approach with delayed-DES and wall-modeled LES capabilities. *International Journal of Heat and Fluid Flow*, 29, 1638–1649.

Simon, D. & Richardson, E. (1963) Forms of bed roughness in alluvial channels. *ASCE Transactions*, 128(1), 284–302.

Simon, F., Deck, S., Guillen, P., Sagaut, P & Merlen, A. (2007) Numerical simulation of compressible mixing layer past an axisymmetric trailing edge. *Journal of Fluid Mechanics*, 591, 215–253.

Simpson, J.E. (1972) Effects of the lower boundary on the head of a gravity current. *Journal of Fluid Mechanics*, 53, 759–768.

Singh, K.M., Sandham, N.D. & Williams, J.J.R. (2007) Numerical simulation of flow over a rough bed. *ASCE, Journal of Hydraulic Engineering*, 133(4), 386–398.

Smagorinsky, J. (1963) General circulation experiments with the primitive equations, I, the basic experiment. *Monthly Weather Review*, 91, 99–165.

Smolarkiewicz, P.K. & Margolin, L.G. (1998) MPDATA: A finite difference solver for geophysical flows. *Journal of Computational Physics*, 140, 459–472.

Sofialidis, D. & Prinos, P. (1998) Compound open-channel flow modeling with nonlinear low-Reynolds k–ε models. *ASCE, Journal of Hydraulic Engineering*, 124(3), 253–262.

Sohankar, A., Davidson, L. & Norberg, C. (2000) Large eddy simulation of flow past a square cylinder: Comparison of different subgrid scale models. ASME, *Journal of Fluids Engineering*, 122, 39–47.

Spalart, P.R. (2000) Trends in turbulence treatments. In: *38th AIAA Aerospace Sciences Meeting and Exhibit, Reno, NV, USA*. AIAA Paper 2000-2306.

Spalart, P.R. (2009) Detached eddy simulation. *Annual Review of Fluid Mechanics*, 41, 181–202.

Spalart, P.R. & Allmaras, S.R. (1994) A one-equation turbulence model for aerodynamic flows. *La Recherche Aerospatiale*, 1, 5–21.

Spalart, P.R., Deck, S., Shur, M.L., Squires, K.D., Strelets, M. & Travin, A. (2006) A new version of detached-eddy simulation, resistant to ambiguous grid densities. *Theoretical and Computational Fluid Dynamics*, 20, 181–195.

Spalart, P.R., Jou, W.H., Strelets, M. & Allmaras, S.R. (1997) Comments on the feasibility of LES for wings, and on a hybrid RANS/LES approach. In: Liu, C. & Liu, Z. (eds.) *Advances in LES/DNS, First AFOSR International Conference on DNS/LES, 4–8 August, Ruston, LA, USA*. Columbus, OH: Greyden Press.

Stacey, M.W. & Bowen, A.J., (1988) The vertical structure of turbidity currents and a necessary condition for self-maintenance. *Journal of Geophysical Research*, 94(C4), 3543–3553.

Stewart, S. (1969) Turbulence, film. Available from: http://www.mit.edu/hml/nctmt.html

Stoesser, T. (2010) Physically realistic roughness closure scheme to simulate turbulent channel flow over rough beds within the framework of LES. *ASCE, Journal of Hydraulic Engineering*, 136, 812–819.

Stoesser, T., Braun, C., Garcia-Villalba, M. & Rodi, W. (2008) Turbulence structures in flow over two-dimensional dunes. *ASCE, Journal of Hydraulic Engineering*, 134(1), 42–55.

Stoesser, T., Fröhlich, J. & Rodi, W. (2003) Identification of coherent flow structures in open-channel flow over rough bed using large eddy simulation. In: *Proceedings of the 30th IAHR, 25–31 August 2003, Thessaloniki, Greece*.

Stoesser, T., Fröhlich, J. & Rodi, W. (2007) Turbulent open-channel flow over a permeable bed. In: *Proceedings of 32nd IAHR Congress, 1–6 July 2007, Venice, Italy*.

Stoesser, T., Kim, S.J. & Diplas, P. (2010a) Turbulent flow through idealized emergent vegetation. *ASCE, Journal of Hydraulic Engineering*, 136, 1003–1017.

Stoesser, T. & Nikora, V.I. (2008) Flow structure over square bars at intermediate submergence: Large eddy simulation study of bar spacing effect. *Acta Geophysica*, 56, 876–893.

Stoesser, T., Palau-Salvador, G. & Rodi W. (2009) Large eddy simulation of turbulent flow through submerged vegetation. *Transport in Porous Media*, 78, 347–365.

Stoesser T., Ruether, N. & Olsen, N.R.B. (2010b) Calculation of primary and secondary flow and boundary shear stresses in a meandering channel. *Advances in Water Resources*, 33(2), 158–170.

Stolz, S. & Adams, N.A. (1999) An approximate deconvolution procedure for large-eddy simulation. *Physics of Fluids*, 11, 1699–1703.

Stolz, S., Adams, N. & Kleiser, L. (2001) An approximate deconvolution model for large-eddy simulation with application to incompressible wall-bounded flows. *Physics of Fluids*, 13, 997–1015.

Stoltz, S., Schlatter, P. & Kleiser, L. (2005) High-pass filtered eddy viscosity models for LES of compressible wall bounded flows. *Journal of Fluids Engineering*, 127, 666–673.

Stoltz, S., Schlatter, P. & Kleiser, L. (2007) LES of sub-harmonic transition in a supersonic boundary layer. *AIAA Journal*, 45(5), 1019–1027.

Stone, B.M. & Shen, H.T. (2002) Hydraulic resistance of flow in channels with cylindrical roughness. *ASCE, Journal of Hydraulic Engineering*, 128(5), 500–506.

Strayer, D.L. (1999) Use of flow refuges by unionid mussel in rivers. *Journal of North American Benthological Society*, 18, 468–476.

Strelets, M. (2001) Detached eddy simulation of massively separated flows. AIAA paper 2001–0879.

Sukhdolov, A.N. & Rhoads, B.L. (2001) Field investigation of three-dimensional flow structure at stream confluences: Part II. Turbulence. *Water Resources Research*, 37(9), 2411–2424.

Tanino, Y. & Nepf, H.M. (2008) Laboratory investigation of mean drag in a random array of rigid, emergent cylinders. *ASCE, Journal of Hydraulic Engineering*, 134(1), 34–41.

Tanino, Y., Nepf, H.M. & Kulis, P.S. (2005) Gravity currents in aquatic canopies, *Water Resources Research*, 41, W12402.

Temmerman, L., Hadziabdic, M, Leschiziner, M.A. & Hanjalic, K. (2005) A hybrid two-layer URANS-LES approach for large eddy simulation at high Reynolds numbers. *International Journal of Heat and Fluid Flow*, 26, 173–190.

Thomas, T.G., Leslie, D.C. & Williams, J.J.R. (1995) Free-surface simulations using a conservative 3D code. *Journal of Computational Physics*, 116, 52–68.

Thomas, T.G. & Williams, J.J.R. (1995) Large eddy simulation of turbulent flow in an asymmetric compound open channel. *Journal of Hydraulic Research*, 33(1), 27–41.

Thompson, J.F., Warsi U.A. & Mastin, C.W. (1985) *Numerical Grid Generation, Foundations and Applications*. North-Holland: Amsterdam.

Tokyay, T. & Constantinescu, S.G. (2006) Validation of a large eddy simulation model to simulate flow in pump intakes of realistic geometry. *ASCE, Journal of Hydraulic Engineering*, 132(12), 1303–1315.

Tokyay, T. Constantinescu, G., Gonzales-Juez, E. & Meiburg, E. (2011b) Gravity currents propagating over periodic arrays of blunt obstacles: Effect of the obstacle size. *Journal of Fluids and Structures*, 27, 798–806. Available from: doi:10.1016/j.jfluidstructs.2011.01.006

Tokyay, T., Constantinescu, G. & Meiburg, E. (2011a) Lock exchange gravity currents with a high volume of release propagating over an array of obstacles. *Journal of Fluid Mechanics*, 672, 570–605.

Tokyay, T., Constantinescu, G. & Meiburg, E. (2012) Tail structure and bed friction velocity distribution of gravity currents propagating over an array of obstacles. *Journal of Fluid Mechanics*, 694, 252–291.

Tominaga, A. & Nezu, I. (1991) Turbulent structure in compound open-channel flows. *ASCE, Journal of Hydraulic Engineering*, 117(1), 21–40.

Touber, E. & Sandham, N.D. (2009) Large-eddy simulation of low-frequency unsteadiness in a turbulent shock-induced separation bubble. *Theoretical and Computational Fluid Dynamics*, 23, 79–107.

Travin, A., Shur, M., Strelets, M. & Spalart, P. (2000) Physical and numerical upgrades in the detached-eddy simulation of complex turbulent flows. In: Friedrich R. & Rodi W. (eds.) *Advances in LES of Complex Flows: Proceedings of the EUROMECH Colloquium 412.* pp. 253–271, Munich: Kluwer.

Uijttewaal, W.S.J. & Booij, R. (2000) Effects of shallowness on the development of free-surface mixing layers. *Physics of Fluids*, 12, 392–402.

Uijttewaal, W.S.J., Lehmann, D. & van Mazijk, A. (2001) Exchange processes between a river and its groyne fields: Model experiments. *ASCE, Journal of Hydraulic Engineering*, 127, 928–936.

Van Balen, W, Blanckaert, K. & Uijttewaal, W.S.J. (2010a) Analysis of the role of turbulence in curved open channel flow at different water depths by means of experiment, LES and RANS, *Journal of Turbulence*, 11, 1–34.

Van Balen W., Uijttewaal W.S.J. & Blanckaert, K. (2010b) Large-eddy simulation of a curved open-channel flow over topography. *Physics of Fluids*, 22, 075108.

Van Balen W., Uijttewaal, W.S.J. & Blanckaert, K. (2009) Large-eddy simulation of a mildly curved open-channel flow, *Journal of Fluid Mechanics*, 630, 413–442.

Van Driest, E.R. (1956) On the turbulent flow near a wall. *Journal of Aeronautical Science*, 23, 1007–1011.

van Prooijen, B.C. & Uijttewaal, W.S.J. (2002) A linear approach for the evolution of large-scale turbulence structures in shallow mixing layers. *Physics of Fluids*, 14(12), 4105–4114.

van Rijn, L.C. (1984) Sediment transport, part III: Bed forms and alluvial roughness. *ASCE, Journal of Hydraulic Engineering*, 110(12), 1733–1754.

Veloudis, I., Yang, Z., Mcguirk, J.J., Page, G.J. & Spencer, A. (2007) Novel implementation and assessment of a digital filter based approach for the generation of LES inlet conditions. *Flow Turbulence and Combustion*, 79, 1–24.

Versteeg, H.K. & Malalasekera, W. (2007) An introduction to computational fluid dynamics: The finite volume method. 2nd edition. Harlow: Pearson Education Limited.

von Terzi, D.A. & Fröhlich, J. (2007) Coupling conditions for LES with downstream RANS for prediction of incompressible turbulent flows. In: *Proceedings of the 5th International Symposium on Turbulence and Shear Flow Phenomena TSFP-5*, 27–29 August 2007, Munich, Germany, Vol. 2. Elsevier. pp. 765–770.

Wang, M. & Moin, P. (2002) Dynamic wall modeling for large eddy simulation of complex turbulent flows. *Physics of Fluids*, 14, 2043–2051.

Weitbrecht, V., Socolofsky, S.A. & Jirka, G.H. (2008) Experiments on mass exchange between groin fields and the main stream in rivers. *ASCE, Journal of Hydraulic Engineering*, 134(2), 173–183.

Wendt, J.F. (1992) *Computational Fluid Dynamics*. Berlin: Springer-Verlag.

Werner, H. & Wengle, H. (1991) Large-eddy simulation of turbulent flow over and around a cube in a plate channel. In: *8th Symposium on Turbulent Shear Flows, 9–11 September 1991, Munich, Germany*. pp. 155–168.

White B. & Nepf H.M. (2008) A vortex-based model of velocity and shear stress in a partially vegetated shallow channel. *Water Resources Research*, 44, W01412.

Wilcox, D.C. (2006) *Turbulence Modeling for CFD*. 3rd edition. La Canada, CA: DCW Industries.

Williams, J.J.R. (2007) Free-surface simulations using an interface-tracking finite-volume method with 3D mesh movement. *Engineering Applications of Computational Fluid Mechanics*, 1, 49–56.

Willmart, W.W. & Lu, S.S. (1972) Structure of Reynolds stress near wall. *Journal of Fluid Mechanics*, 55, 65–92.

Xiao, X., Edwards, J.R. & Hassan, S.A. (2004) Blending functions in hybrid large eddy/Reynolds-averaged Navier Stokes simulations. *AIAA Journal*, 42(12), 2508–2515.

Xie, Z.T. & Castro, I.P. (2008) Efficient generation of inflow conditions for large eddy simulation of street-scale flows. *Flow Turbulence and Combustion*, 81, 449–470.

Yang, K.S. & Ferziger, J.H. (1993) Large eddy simulation of turbulent obstacle flow using a dynamic subgrid-scale model. *AIAA Journal*, 31, 1406–1413.

Yoon, J.Y. & Patel, V.C. (1996) Numerical model of turbulent flow over sand dune. *ASCE, Journal of Hydraulic Engineering*, 122, 10–18.

Yoshizawa, A. (1982) A statistically-derived subgrid model for the large-eddy simulation of turbulence. *Physics of Fluids*, 25, 1532–1539.

Yoshizawa, A. & Horiuti, K. (1985) A statistically-derived subgrid-scale kinetic energy model for the large-eddy simulation of turbulent flows. *Journal of the Physical Society of Japan*, 54, 2834–2839.

Youngs, D.L. (1982) Time-dependent multi-material flow with large fluid distortion. In: Morton K.W. & Bianes M.J. (eds.), *Numerical Methods for Fluid Dynamics*. New York: Academic Press.

Yue, W.S., Lin, C.L. & Patel, V.C. (2005a) Coherent structures in open-channel flows over a fixed dune. *Journal of Fluids Engineering-Transactions of the ASME*, 127, 858–864.

Yue, W.S., Lin, C.L. & Patel, V.C. (2005b) Large eddy simulation of turbulent open-channel flow with free surface simulated by level set method. *Physics of Fluids*, 17, 025108.

Yue, W.S., Lin, C.L. & Patel, V.C. (2006) Large-eddy simulation of turbulent flow over a fixed two-dimensional dune. *ASCE, Journal of Hydraulic Engineering*, 132, 643–651.

Yuksel-Ozan, A., Constantinescu, G. & Nepf, H.M. (2012) An LES study of exchange flow between open water and a region containing a vegetated surface layer. In: *CD-ROM Proceedings of 3rd IAHR International Symposium on Shallow Flows, June 2012, Iowa City, IA, USA*.

Zang, Y., Street, R.L. & Koseff, J.R. (1993) A dynamic mixed subgrid-scale model and its application to turbulent recirculating flows. *Physics of Fluids*, 5, 3186–3196.

Zedler, E.A. & Street, R.L. (2001) Large-eddy simulation of sediment transport: Currents over ripples. *ASCE, Journal of Hydraulic Engineering*, 127, 444–452.

Zeng, J., Constantinescu, G., Blanckaert, K. & Weber, L. (2008) Flow and bathymetry in sharp open-channel bends: Experiments and predictions. *Water Resources Research*, 44(9), W09401.

Zhang, J.F. & Dalton, C. (1997) Interaction of a steady approach flow and a circular cylinder undergoing forced oscillation. *ASME, Journal of Fluids Engineering*, 119, 802–813.

Zhou, J., Adrian, R.J., Balachandar, S. & Kendall, T.M. (1999) Mechanisms for generating coherent packets of hairpin vortices in channel flow. *Journal of Fluid Mechanics*, 387, 353–396.

Zhu. J. (1991) A low-diffusive and oscillation-free convection scheme. *Communications in Applied Numerical Methods*, 7, 225–232.

Index

T - #0513 - 071024 - C266 - 245/170/12 - PB - 9780367576387 - Gloss Lamination